and then you're dead

PENGUIN BOOKS

AND THEN YOU'RE DEAD

CODY CASSIDY has worked as the sports editor for Zimbio.com, a sports reporter for *Stanford Athletics*, and a writer for *Coach* magazine. He has no firsthand experience with any of the scenarios described in this book.

PAUL DOHERTY is codirector and senior staff scientist at San Francisco's famed Exploratorium Museum. He has cowritten numerous books, including *The Exploratorium Science Snackbook*, *Explorabook*, and the *Klutz Book of Magnetic Magic*. He received his PhD in solid state physics from MIT.

And Then You're Dead

What Really Happens

 If You Get Swallowed by a Whale,

 Are Shot from a Cannon,

 or Go Barreling over Niagara

Cody Cassidy and
Paul Doherty, PhD

PENGUIN BOOKS

PENGUIN BOOKS

An imprint of Penguin Random House LLC
375 Hudson Street
New York, New York 10014
penguin.com

Illustrations by Cody Cassidy

ISBN 9780143108443

Printed in the United States of America
10 9 8 7 6 5 4 3 2 1

Set in Scala OT
DESIGNED BY KATY RIEGEL

Cody:

To Mom and Dad

Paul:

*To Professor Paul Tipler,
who showed me how to inspire students
to learn science by making it interesting,
relevant, fun, and correct*

Contents

What Would Happen If . . .

Introduction

Be honest. When you are reading a random obituary do you sometimes find yourself skipping to the bottom, searching for the cause of death, only to be frustrated by the lack of an explanation or a maddeningly vague "death by fluke accident"? Did the poor sap freeze while ice swimming? Was he squished by an asteroid or was he swallowed by a whale? Sometimes they won't even tell you!

And when they do reveal a cause of death—say the obituary provides a tantalizing detail like "tragically killed by an oversize magnet"—the story quickly moves on to next of kin while you're left wondering if magnetism even *can* be lethal. They are skipping the most interesting part!

We understand your frustration, so we set out to resolve it. We pick up where even the most elucidating obituary leaves off.

We tell you what *really* happens when you jump into space wearing only shorts and a T-shirt. We explain why Boeing doesn't let you roll your window down on the 747, and we explore the problems with swimming in the deepest part of the ocean with as much science and gruesome detail as your stomach will allow.

In other words: Stephen King meets Stephen Hawking.

The upside in wading through all this gruesomeness is you may accidentally learn some science, a bit of medicine, and what to do if a shark begins circling you (encourage him to eat your entire leg—not just a chunk).

How did we get our answers?

When we could we used the experiences (or autopsies) of daredevils (or the unlucky) to figure out what actually happens when you go over Niagara Falls in a barrel, stick your hand in a particle accelerator, or get stung in the testicle.

For some of the scenarios there weren't firsthand accounts. So far nobody has actually jumped into a black hole, taken a bath in the world's coldest tub, or dug a hole to China and leapt into it.

To get answers to these questions we used military studies (thank you, 1950s-era U.S. Air Force, for subjecting real people to life-threatening experiments), medical journals, astrophysicists' hypotheses, and the research of professors curious about the slipperiness of banana peels.

Sometimes our answers took us to the edge of human knowledge. If this book were written just twenty years ago we would have sworn that, at least in this universe, you could not die from an oversize kitchen magnet. Fortunately, we didn't write it back then because you absolutely can and it's glorious.

Because we were often reaching the frontiers of science in

search of gruesome deaths, we also relied on speculation—the most science-based, as-accurate-as-we-think-anyone-could-get speculation. But it's still speculation.

Meaning if you try one of these scenarios, say, if you sky-dive from the space station, swan dive into a black hole, or leap into a volcano, and your experience does not mimic what we have described or, worst of all, you don't even die, we sincerely apologize.

Send us a note and we will amend our second edition.

And Then You're Dead

What Would Happen If . . .
You Were in an Airplane
and Your Window Popped Out?

Like most people who have traveled in a modern airplane, you have probably spent a good bit of time staring out the window at the lovely clouds, sunsets, and beautiful views. And, like most people, you have probably wondered, what happens if this thing pops out?

The answer depends on your altitude. If you were within the first few minutes of flight and still under 20,000 feet, you would probably be okay. You could still breathe for a half hour before you passed out at that altitude, and the pressure difference wouldn't be great enough to suck you out. It would be a little chilly, but as long as you're wearing a sweatshirt you should be fine.

It would also be noisy. The wind blowing past your open window would turn the plane into the world's largest flute, so getting the attention of a flight attendant would be a problem.

All in all, though, not bad, and a lot better than if the window popped out at a cruising altitude of 35,000 feet.

The air inside a plane's cabin is pressurized to around 7,000 feet because of the whole breathing thing. If you're at 35,000 feet and the window pops out, the plane rapidly depressurizes, and that leads to some issues.

The first thing you would notice is all the air getting sucked out of every orifice in your body. And because it's humid air, it would condense and come out as a fog. That would happen to everybody, so the entire plane would be a thick fog of everyone's body air. Gross.

Fortunately that would clear up in a few seconds, because the air in the plane is getting sucked out of the open window. Unfortunately, it's not a neighbor's window, it's yours, and that makes a big difference.

If you were sitting just two seats away from the missing window, the wind would be rushing out of the plane with hurricane speed, but that's still slow enough that if you were wearing a seat belt you would be held fast. Unfortunately, you chose the window seat, where the air would rush out at 300 miles per hour—fast enough to pull you up and out of your seat even if you're strapped in. (One of the less-mentioned cons of choosing the window over the aisle.)*

Another reason your friend in the aisle seat would be saved is because airplane windows are smaller in diameter

---

*Why do a few feet make such a difference? Picture it this way: When you plug up your bathtub, the power of the water sucking the plug into place gets exponentially greater the closer it gets. Same thing when it comes to airplane windows, and you're the plug.

than your shoulders. According to research by Harvard University on the human body, the average American has 18-inch-wide shoulders, and the Boeing 747 aircraft's windows are only 15.3 inches tall—so you would not be sucked all the way out of the plane, just partway.* That's good for everyone in the plane. It would save you from a long fall, for one, and for everyone else your body would serve as a decent plug. It would slow down the air's escape from the plane and give people more time to put on their oxygen masks.

Your troubles, on the other hand, would only be beginning.

The first thing you might notice about your new environment would be the wind. The 600-miles-per-hour gale blasting you in the face would push you against the aircraft, wrapping you in a J-shaped figure around the side of the plane.†

The second thing you would notice would be the cold. The temperature at 35,000 feet is 65 degrees below zero. In that chill your nose would become frostbitten within a few seconds.

The third issue is not something you would notice but is probably the most life-threatening. In addition to the abrupt drop in temperature, there would be a more serious change in air pressure. At 35,000 feet the air is so thin you wouldn't get enough oxygen molecules per breath to survive, only you would not know you were suffocating. Your body cannot

---

*This is where real life differs from the James Bond movie *Goldfinger*. Goldfinger would not have been sucked out of the window; he just would have been stuffed into it.
†Instead of being pressed into the plane, you would bang against it because of something called reverberation dynamics, which is the same principle that explains a flag flapping in the wind instead of being held in one position. Even if it seems like the wind is constant, it isn't, and the flag is in a perpetual state of change and adjustment. Your changes and adjustments would be your face slamming against the aircraft repeatedly.

detect when there's too little oxygen; the only thing that gives you that running-out-of-breath feeling is too much carbon dioxide in your blood. So you would keep breathing like everything was fine, but it wouldn't be. You would have less than fifteen seconds of consciousness before you passed out—and four minutes before brain death.

That goes for people inside the plane as well. As soon as your window popped out they would have fifteen seconds to put on their masks before they passed out—maybe a bit more if your upper body formed a good seal on the window—and really only eight seconds before their brains became so oxygen starved they would be too confused to put on their masks.*

So to recap, you would be halfway out of the airplane, your face would be slamming against the side of the plane, you would have frostbite, and you would be on your way to unconsciousness. But you wouldn't be dead yet and, surprisingly, if the pilot acted quickly and got down below 20,000 feet within four minutes, you might survive the experience. We know this because it's happened.

Captain Tim Lancaster was climbing past 20,000 feet in his British Airways flight in 1990 when the front windscreen popped off. He was immediately sucked out of his seat belt and out the window. Everything loose in the cockpit flew out and the flight door jammed into the controls, sending the plane into a steep dive. Nigel Ogden, a flight attendant who happened to

---

*This happened on professional golfer Payne Stewart's private jet in 1999. His plane decompressed at 30,000 feet and the pilots weren't able to put their masks on in time. Because the plane was on autopilot when it depressurized, it continued flying for 1,500 miles before it ran out of fuel and crashed in South Dakota.

be in the cockpit, managed to grab the pilot on his way out and reported the following to the *Sydney Morning Herald*:

> Everything was being sucked out of the aircraft: even an oxygen bottle that had been bolted down went flying and nearly knocked my head off. I was holding on for grim death but I could feel myself being sucked out, too. John rushed in behind me and saw me disappearing, so he grabbed my trouser belt to stop me slipping further, then wrapped the captain's shoulder strap around me . . .
>
> I thought I was going to lose him, but he ended up bent in a U-shape around the windows. His face was banging against the window with blood coming out of his nose . . . and his arms were flailing.

Eighteen minutes after losing the windscreen the copilot managed to land the aircraft, with his pilot staring at him from the other side of the window the entire time.

Somehow, after firefighters managed to extract the pilot from his awkward position, he survived with only frostbite and a few broken ribs.

Because of the smaller window, you may not need to rely on heroics from your fellow passengers—with just quick action from your pilot, you could enjoy an uncomfortable but scenic trip down.

What Would Happen If . . .
You Were Attacked
by a Great White Shark?

Like all predators, sharks are not interested in fair fights. Even for the winners, fair fights lead to injuries, and injuries mean a slow and hungry animal. So predators prefer devastating blowouts with as little risk as possible, which makes you the perfect opponent: You're slow, weak, and completely oblivious in the water. Fortunately, you don't taste very good. You're the squirrel of the ocean, too much bone and not enough fat. Still, sharks are curious creatures and attacks happen—usually from the smaller species that aren't as dangerous.

But not always. Big sharks *can* attack. The great white can grow to twenty feet, and even its exploratory nibbles are devastating. Why might the shark go for a bite?

It probably would not be for food. Researchers have stitched shark victims back together and discovered not a single morsel missing. When great white sharks bite a human, they are like children scrambling peas on their plate. Careful reconstruction reveals nary a pea eaten. We must taste so terrible to sharks that, frankly, we should be a little insulted.

So if we taste so horrible, why bite us at all? One popular explanation is that it's a case of mistaken identity. The theory goes that sharks mistake human swimmers for normal seal prey and take a bite, only then realizing their error and spitting the person out like a diner mistaking the salt for the sugar. It is plausible, but there is little science to back up this theory. There are visual similarities between a surfer and a seal from a shark's point of view, but that does not explain important differences in the way a shark attacks a swimmer versus the way it strikes a seal.

Researchers placed dummies in chummed water to observe the way sharks approached them. Unlike seal attacks, in which the shark comes from below and hits the animal with one devastating surprise attack, the sharks swam in circles around the dummies—checking them out with multiple passes before striking. The nature of the bite was also a more exploratory, open-bite slash as opposed to the full-gusto chomping bite a shark uses with a seal—like the difference in how you approach a carton of fresh milk as opposed to one close to its expiration date.

So far the evidence suggests that it is not confusion at work when a great white shark attacks, but mere curiosity. Sharks can sense movement by detecting small changes in water pressure, and swimmers are moving, particularly if they have just spotted a fin. This motion can pique a great white's interest, and sharks seem to operate under a "when in doubt, bite it" policy.*

Incidentally, this is common behavior for many predators—if you have a cat you may have seen this explore-the-world-via-biting behavior. But exploratory biting by sharks significantly differs from your cat's. There aren't any reliable measurements of exactly how strong a great white's bite is, but the few experiments that have been done all come to the same general conclusion: It's strong enough. In at least one instance a great white bit a man in half as clean as any guillotine.

So let's say you're splashing about in the waves and, unbeknownst to you, you attract the attention of a curious great white.

First of all, you would have every right to be upset. Not because you could be slashed to death in a moment, but because the odds of this happening are *infinitesimal*. If you're headed for a day at the beach, you're ten times more likely to fall down your stairs and die on your way to your car. Once you get in your car you're way more likely to die in an accident driving to the beach, and once you get to the beach you're far more likely to die in a collapsing sand pit on your way to the

---

*It's important to note that we're talking about great white sharks here—which kill the most people but don't appear to do it out of hunger. Another breed of shark, called the oceanic whitetip, has intentionally killed and eaten humans. However, attacks from whitetips are uncommon (usually survivors of shipwrecks) because they frequent open ocean, far away from people, whereas great whites often patrol beaches.

The most famous oceanic whitetip attack occurred in 1945 just before Japan's surrender when a navy ship, the USS *Indianapolis*, was torpedoed near the Philippines. Nine hundred men hit the water alive, but because of a miscommunication they weren't rescued for four days. Oceanic whitetip sharks, attracted to all the commotion, began feeding on the sailors. By the time the survivors were rescued, the sharks had killed and eaten as many as 150 men.

water. And even if you avoid those sand pits and make it to the waves, you face the greatest threat of all: drowning. Once you hit the waves, you're a hundred times more likely to drown than die from a shark attack.

But let's say you're lucky and dodge all these bullets. And then you get really unlucky and a great white decides to go for a nibble.

Sharks like to attack from below and behind, so you would probably be struck in the legs. They also have bad table manners: They don't chew. They tear and rip by thrashing their heads from side to side and rolling their bodies. From spiral teeth markings on bone we can see that sharks like to saw flesh off and then swallow it whole.

The good news is that 70 percent of attacks are one bite only. The bad news is that a single bite and rip from a great white shark is more than enough to remove your leg. However, that can actually work *for* you.

The great danger in a leg chomping is a cut to your femoral artery. In general, injuries to arteries are more dangerous than those to veins because arteries carry blood *from* your heart and are under pressure, so when they're severed they squirt—as opposed to veins, which just drool.

The femoral is one of the worst arteries to sever. It's responsible for oxygenating your entire leg, and nearly 5 percent of your blood volume passes through it every minute.

Exactly how the shark bites your leg would determine whether you have any chance at all. The human body cannot afford to lose 5 percent of its blood volume per minute—that equates to death in four minutes—so you would think that if your femoral artery was severed, your story would be a short one. But that's not always the case.

Right now, as you read these words, your femoral artery is under a small bit of tension, like a stretched rubber band. If it were severed *cleanly* by the shark, it would snap back into the stump of your leg, where your muscles could pinch it shut—slowing the leak and giving you time to get a tourniquet on. But if it were slashed *unevenly,* or at an angle, it wouldn't recede correctly—that's bad. You would black out in thirty seconds. From there you would go into circulatory shock—a deadly positive feedback loop wherein your tissues die from lack of blood, swell up, and compound the problem by blocking blood flow elsewhere in the body.

Four minutes after the attack, if your femoral was cut unevenly, you would have lost 20 percent of your blood and you would enter a critical stage. Your heart needs a minimum blood pressure to keep beating, and once you lost 20 percent of your blood volume you would drop below that threshold. After that it would only be a few minutes until complete brain death.

All of this assumes you were lucky and the shark did the expected and attacked from behind. A frontal attack on your head and torso is less likely but worse. Losing your head is bad because, one, your brain is in it and, two, tourniquets are far less effective on your neck than they are on your legs (for details, see Wikipedia for "Hanging").

Lawyer's note: Seriously—do not put a tourniquet around your neck.

What Would Happen If...
You Slipped on a Banana Peel?

I<small>F YOU SEE</small> a banana peel on the floor, how concerned should you be? If the cartoons are to be believed, the answer is, of course, very. Cartoons might understate banana peel danger by overstating the strength of your skull, but the cartoons aren't kidding about the slipperiness of banana peels. Rigorous scientific study has confirmed bananas as the most dangerous of all fruit peels.

Slipperiness is measured by placing a block of a given material on a ramp of another material and then slowly increasing the angle of the ramp. The tangent of the angle of the ramp when the object starts to slide gives the coefficient of friction (CoF), and it usually scales from 0 (the slipperiest) to 1 (stickiest), though in some stickier situations it can go as high as 4.[*] Rubber on a cement sidewalk has a near slip-proof CoF of 1.04.

---

[*] A CoF larger than 1 means the object slips at an angle greater than 45 degrees. The highest CoF we can find is the rubber on the tires of top fuel dragsters, which when spinning have a CoF on pavement of 4 (they could climb a 75-degree wall).

Then there's the other end of the spectrum. Sliding on socks across a wooden floor has a CoF of only 0.23, and ice is even slipperier. A walk across an ice rink can have embarrassing consequences because rubber on ice registers a potentially painful CoF of 0.15.*

Banana peels put all that to shame.

We know this thanks to a few daring professors at Kitasato University in Minato, Japan, who decided to double-check the cartoons. Dr. Kiyoshi Mabuchi and his team peeled a bunch of bananas, threw them on a wooden floor, and stepped on them with rubber-soled shoes (hopefully they had a spotter). Then they measured the forces involved.

It turns out Elmer Fudd might not have been as clumsy as we all thought. Banana peels on wood have a CoF of only 0.07—twice as slippery as ice and five times slipperier than wood. Mabuchi and his team of researchers weren't done, though. Was the banana peel slippery merely because of its water content? Would other fruit peels result in similar slippage?

To find out they peeled apples and tangerines and ran the same rigorous experiment: They stepped on them. The apple peel came in a distant second, at 0.1, and the tangerine peel was by far the stickiest, with a CoF of 0.225 (about the same as stepping on a wooden floor *without* a peel).

So if you're walking through a fruit factory and have a choice

---

*Lubricated surfaces have even less friction. The synovial fluid that lubricates your joints, for example, is one of the slipperiest substances in the world—registering at a CoF of .0003, which is a good thing, otherwise it would give cracking your knuckles a more literal interpretation.

of peels to step on, remember this: It's not just a joke; banana peels are the worst. Under pressure, a banana peel oozes a gel that turns out to be extremely slippery. Your foot and body weight provide the pressure. The gel provides the humor.

Why is slipperiness so important? Walking is really just a series of falls and catches. With each step you fall forward, and with the next one you catch yourself and begin the process over again. Banana peels mess up the catching part. If you just stand on a slippery surface, you will probably be okay. But if you take a step, you initiate a fall. To stop it, your leading foot hits the ground with forward momentum at a strike angle of 15 degrees. If you know you're walking on a slippery substance, you will change your gait to decrease that angle, demand less friction from the floor, and lessen your chances of taking a tumble. Stray banana peels have a way of sneaking up on you, though, and research suggests that taking a normal step on a substance with a CoF of less than 0.1 results in a fall 90 percent of the time.

Of course, the real danger with falling is injuring your brain, an essential organ that lives high off the ground. Learning to walk upright sometime 4 to 6 million years ago was a big advancement for the human species, but it did introduce the problem of a slip-and-fall. If you were, say, the height of a small dog and you fell, your head would not build up enough speed to do any damage when it hit the sidewalk.* You could dance on banana peels, because the difference between falling twelve inches and hitting your head and falling

---

*This is where bugs really have us beat. No bug in the history of bugs has ever fallen to its death.

six feet on the same organ is the difference between a bruise and a broken skull.

The force generated by an unrestrained falling adult onto something solid is more than enough to crack a skull. In ballpark terms (everyone's head is a little different) your skull would crack with as little as an unrestrained three-foot fall onto a hard surface. The skull is stronger in the front and back, and weaker on the sides, but even if you fall onto the stronger frontal bone, a fall of six feet is enough to crack it—especially if you pitch forward.

Either way, if you cannot protect your head from a fall of six feet, your skull would fracture. Fractures are dangerous for a few reasons, but bleeding is the big one. Your brain is a blood hog, which means cracking it results in a lot of bleeding inside, putting you in immediate and deep trouble.

Bleeding inside your skull can be far more dangerous than bleeding anywhere else. And it's not just because you can bandage a leg wound and you can't an internal skull bleed. It's because your skull is a solid container carrying fragile cargo. If your head starts filling with blood, your brain gets squeezed. Too much blood within your skull creates pressure that strangles the rest of your brain and chokes off and kills critical brain functions, like remembering to breathe.

Of course your brain knows how fragile it is, and if you slip it works very hard to put something in the way to break your fall—hands, elbows, knees—anything but itself. Which is why you see more bruised butts than broken heads and why banana peels are usually funny, not lethal.

But "usually" isn't the same as "always." And that brings us to Mr. Bobby Leach, the English daredevil of Niagara Falls.

Since 1901, roughly fifteen people have attempted to go over Niagara Falls for the fame or the thrill (see p. 57 for what happened when they did). Five of them drowned; most never went back. ("I'd rather stand in front of a cannon and be blown to death," responded the first survivor, "than do that again.")

But Bobby Leach was a professional stuntman, daredevil, and circus performer who cheated death for a living. In 1906, he climbed into a steel barrel and went over the falls. He survived, although he needed six months of hospitalization to recover from two wrecked knees and a broken jaw.

Afterward, he went on to a successful lecturing career, touring the world with his barrel and posing for photos. In 1926, he was in New Zealand when he slipped on an unidentified fruit peel on a sidewalk in Auckland and gashed his leg. A few days later, Bobby Leach died from the complications.

What Would Happen If . . .
You Were Buried Alive?

You can measure your pulse by putting two fingers on your jugular vein in the crook between your chin and neck. In a minute you should count around seventy beats. If the count is lower than twenty-six, you should finish this chapter in the back of an ambulance.

If you cannot feel anything, your finger is probably in the wrong place, but even if it isn't you're not necessarily dead. Sometimes a pulse is so weak it cannot be felt.* This posed a problem for doctors in the Middle Ages, when feeling for a

---

*Maybe you suffer from sleep paralysis. During parts of sleep the body is paralyzed, which is fine unless the brain makes a mistake and you wake up during this paralysis and your muscles don't turn on. On average this happens to everyone once in their life and it usually lasts less than a minute, but in some cases can last up to an hour and can really confuse EMTs. In one case a woman made it all the way to the morgue before she woke up.

pulse was the only way to determine if a patient was alive.* Occasionally comatose patients were declared dead, only to wake up in the morgue sometime later. Soon, concerned people asked to be buried with a bell above their grave and a string running into their coffin, just in case.†

Doctors today have more sophisticated means of deciding whether you're dead (they look for electrical signals from your heart and brain). But let's say your physician has an early dinner reservation and cuts a few corners. He signs your death certificate, grabs his coat, and jumps in a cab, headed for dinner and a show. You, meanwhile, are in a gurney being wheeled down to the loading dock and then placed in the back of an ambulance, headed for the morgue and a hole in the ground. What would happen next?

Once you're placed in the airtight coffin you would start using up its oxygen. A typical coffin has 900 liters of air and you take up 80 of it, so you would have 820 left. Your lungs take in a half liter per breath, but you use up only 20 percent of the oxygen per breath, meaning you could rebreathe the same air a few times before completely depleting it.

Of course, you wouldn't need to breathe every last bit of oxygen before running into trouble. Air is 21 percent oxygen and that's where you're happy. Once you began using up oxygen you would quickly run into issues. Breathing air with 12

---

*Another test: Doctors would hold up a mirror near your mouth, and if you were breathing, your humid exhalation would "fog the mirror." Thus the origin of the phrase "Anyone who could fog a mirror could do this job."
†Edgar Allan Poe was one of them. He had a thing about being buried alive.

percent oxygen would give you headaches, dizziness, nausea, and confusion as your brain cells began to starve.

Your coffin has enough oxygen to last around six hours before you start to asphyxiate—as long as you stay calm. You would think that you would last longer holding your breath, but that actually increases your oxygen usage when your body overcompensates for the CO_2 buildup with bigger breaths than it needs. Slow, controlled breathing is the way to go.

Once the oxygen drops to 10 percent you would go unconscious without warning and quickly fall into a coma.* Sudden death happens at 6 to 8 percent oxygen.

But here's where it gets interesting and a little complicated. There's another issue competing to kill you. By breathing, you are replacing the oxygen in your coffin with CO_2.† That's a problem.

The excess CO_2 you are breathing binds with your blood and limits the amount of oxygen it can carry into your tissues—effectively asphyxiating your vital organs. Air with 0.035 percent CO_2 is normal, but in your airtight coffin that percentage rises quickly. Once the CO_2 level rises to 20 percent it will render you unconscious in two to three breaths and can kill you within minutes.

Along the way it will also poison your central nervous system, which would manifest as confusion and delirium—so perhaps you would see a ghost in your coffin?

---

*Q: What if you were buried with a few potted plants—would that help? A: Sadly, no, they don't create oxygen quickly enough to make up for the amount of space they take up.

†This was the same issue the astronauts on *Apollo 13* faced after they were forced to move to the lunar excursion module.

Between the increasing CO_2 levels and decreasing oxygen, it's a close race to kill you, but in the end you will be poisoned to death by your own exhales first. The CO_2 level will rise to lethal levels in only 150 minutes, killing you two hours before your coffin ran out of oxygen.

It could be worse, though, if your grave diggers were really in a hurry and skipped the whole coffin part. That might sound like a better alternative—maybe you think you could escape? However, in reality you would die far faster.

Under six feet of dirt you might as well be encased in cement. Six feet of dirt weighs about five hundred pounds on your chest. In other words: You are not getting out. Regardless of any zombie movie you might have seen, if you ever see an empty grave you can be sure it was an outside job.

But some good news: You would not immediately suffocate. Most of your muscles are too weak to lift five hundred pounds, but your diaphragm isn't—which is important. You need it to lift the dirt and allow your lungs to inflate. So you could still *physically* breathe. Unfortunately, there would not be much *to* breathe.

In snow avalanches, which resemble dirt burials, victims who live through the initial slide but are buried under snow have a predictable survival pattern: Every hour the survival rate drops in half, so if you are buried for an hour, your chances are 50 percent; two hours, 25 percent; and so on. Those survival times would probably look even worse in dirt burials, because fresh snow is 90 percent air while dirt is mostly just dirt. Either way, in ice or dirt, forming an air pocket with your arm is key.

Of course if you're worried about being buried alive, fear

not. You would die long before you made it to the grave. Even if you have a lazy doctor, trips to the morgue are fatal. Before you were buried they would give you the world's worst blood transfusion. To preserve your tissues, morticians replace your blood with formaldehyde, which is, sadly, or perhaps mercifully, fatal.

What Would Happen If . . .
You Were Attacked
by a Swarm of Bees?

MICHAEL SMITH WAS tending to his hive when one particularly adventurous bee flew up his shorts and stung him in the testicle.

Surprisingly, it did not hurt as much as he had feared—which sparked a question: If that is not the worst place to be stung, what is?

Shockingly, he discovered that no one had ever volunteered to intentionally sting themselves the hundreds of times necessary to get a firm answer.

Michael Smith had found a new calling and a new daily routine.

Five times every morning—always between the hours of nine and ten—he would carefully hold a bee with forceps and press it against his skin until it stung him. The first and last stings were always to his forearm and were used as a control, an automatic 5 on his 1–10 pain scale. The middle three stings were located on whichever unlucky body part he had chosen that morning to test. In all, he tested twenty-five different spots over three months. And to answer your question, this is a man who had already been stung on the testicle, so, yes, he tested that *other* body part as well.

It turns out that the least painful places to be stung are the skull, middle toe, and upper arm—they all registered a paltry 2.3 on Smith's pain scale, followed closely by the buttocks, which scored a slightly higher 3.7.

At the other end of the spectrum are the face, penis, and inside of the nose.

Smith discovered that the people who talk about the fine line between pleasure and pain haven't spent much time with bees on their privates. "There is definitely no crossing of the wires between pleasure and pain down there," Smith told *National Geographic*. Although if he were forced to choose, Smith reports he would rather attend his bees without drawers than without a mask. Neither, though, he adds, would make him happy.

"Stings to the inside of the nostril were especially violent, electric, pulsating," says Smith, "and immediately induced sneezing, tears, and a copious flow of mucus."

The final determination? According to Smith (and Smith alone, though he welcomes a larger sample size if you're interested) the penis is a 7.3, the upper lip is an 8.7, and the very worst place to be stung? Inside the nose, an even 9.0.

Little known fact: One bee sting begets others. When a honeybee stings you it simultaneously releases a pheromone cocktail that lets the hive know it needs defending. The dominant ingredient in this pheromone, incidentally, is something called isoamyl acetate, which is a common ingredient in certain kinds of candy because it tastes like bananas. It's also used in Hefeweizen beer. In other words, don't eat banana-flavored Runts or drink a wheat Bavarian beer before rummaging around in beehives.

If you ignored this advice, you would alarm the hive and peeved bees would fly to the rescue. Stingers are barbed, so when the bees flew away—or tried to—their stingers would stay, disemboweling them and making honeybees natural kamikazes.*

Even after a stinger is disconnected, though, it works its barb back and forth to dig itself deeper into your nose, all the while pumping its toxin from a sac in the base of the stinger into your flesh.

A bee sting's venom works in much the same way that all insect poisons work—by hacking into your cells and changing chemical reactions to produce the results it wants.

In your case, the bee's venom penetrates your cell's membranes using a chemical called melittin. The melittin has a cellular bomb in its backpack in the form of phospholipase A_2. If the target is a blood cell it will be destroyed, and if it's a neuron it will misfire—interpreted by your brain as jolts of pain.

Still more chemicals go to work on other bodily functions.

---

*Due to the fact that a sting is fatal to a honeybee, they reserve it for their larger predators. For the smaller ones, like the Asian giant hornet (which has a real sweet tooth for their honey), they have a rather unique killing method. The honeybees surround the intruder in a tightly packed ball and use a combination of their body heat and carbon dioxide emissions to overheat and asphyxiate the thief.

One restricts blood flow, preventing your body from diluting the toxin, which is why the pain persists, while another builds a sort of chemical bridge within your tissues, allowing the toxins to spread and target new cells.

You may interpret the experience as a 9 on the bee-sting pain scale, but it's middle of the road when it comes to insect stings. Which brings us to the other leading authority on the subject, the poet of pain, Justin O. Schmidt.

On the Justin O. Schmidt sting pain index, a bee sting ranks as a mere 2 out of 4. Schmidt knows of what he speaks: He's allowed himself to be stung by more than 150 insect species, turning him into a connoisseur of pain and allowing him to create the world's first all-insect-sting pain scale.

At the low end of the rating scale, clocking in at a mere 1.0, is the sweat bee—with a sting described by Schmidt as "light, ephemeral, almost fruity. A tiny spark has singed a single hair on your arm."

The honeybee, yellow jacket, and bald-faced hornet are all 2s. The bald-faced hornet's sting, if you haven't had the pleasure, feels "rich, hearty, slightly crunchy. Similar to getting your hand mashed in a revolving door."

The yellow jacket is "hot and smoky, almost irreverent. Imagine W. C. Fields extinguishing a cigar on your tongue."

Among the stings that rank above the yellow jacket is the red harvester ant, found in the southwest United States (different from the fire ant), whose sting is a 3 and feels "bold and unrelenting. Somebody is using a drill to excavate your ingrown toenail."

One of the fiercest stings in the insect world comes from

the tarantula hawk. It's found all over the world, including the United States, and it rarely stings people.* But if you should be so unfortunate, its sting is "blinding, fierce, shockingly electric. A running hair dryer has been dropped into your bubble bath."

The title of world's worst sting belongs to that of the legendary bullet ant, found in the tropics of Central and South America. It bests the tarantula hawk's sting not only with its intensity but with its endurance as well.

According to Schmidt, the bullet ant provides a "pure, intense, brilliant pain. Like fire-walking over flaming charcoal with a 3-inch rusty nail in your heel."

The bullet ant may have the most painful sting, but because it doesn't attack in large numbers, it isn't the most dangerous. That distinction falls to the honeybee. A lethal dose of honeybee stings is eight to ten bee stings per pound of body weight.

Since each honeybee can sting you only once, if you're 180 pounds you would need something like 1,500 bee stings to get a heart-stopping dose of their nerve toxin (assuming, of course, you aren't allergic, in which case it may take only one).

Note, however, that 1,500 is only a guideline. There have been outliers. Some have been stung by more and survived. In one notable instance a man survived despite doctors finding

---

*It does sting and paralyze tarantulas, after which it lays an egg on the spider. When the egg hatches, the larva eats its way inside the tarantula and begins consuming it, though the larva takes care to avoid the major organs of the spider so as to keep it alive for as long as possible. When the young wasp is ready, it finally bursts out of its abdomen, like the alien out of Kane's stomach. And now you somehow feel sorry for tarantulas.

more than 2,200 stingers in his body. He was swarmed so aggressively that he dived under water. Unfortunately, the cloud of bees continued hovering over him and was so dense that he was forced to swallow bees in order to get a breath when he surfaced.

He survived, probably because the stings were spread out over some minutes, but by the time the bees decided he was sufficiently punished, his face was black with stingers.

No word on where that ranks on the Smith index.

What Would Happen If . . .
You Were Hit by a Meteorite?

THE NEXT TIME you are stargazing, keep an eye out for the brightest objects in the sky. Excluding the moon, the brightest "star" you see shouldn't be a star at all but the planet Venus. If you see something brighter, keep watching. You might have a problem. If the object gets brighter than the moon and then brighter than the sun, then you definitely have a problem—a meteorite* is headed straight at you. At that point, there's no ducking or covering, so you might as well sit back and enjoy the show.

Let's say the speeding space rock you are standing under is one mile wide. That means that even though its devastation would be enormous, it would not be a planet killer.

From your perspective the meteor would look like a star

---

*The verbage on meteors is confusing. Here's how it works: A meteor is the flash of light in the sky. A meteorite is the solid object that made the flash and reached the ground. A meteoroid is the solid object before it hits the atmosphere.

that kept getting brighter. First it would shine brighter than the sky's brightest star (Sirius), then it would outshine Venus, then it would become even brighter than the moon—and then, surprisingly, you would die.

"Surprisingly" because you might expect to live a few more seconds than you actually would. You might expect to be squashed, but you would actually die some tens of seconds before the rock hit you.

As the meteorite plummeted toward Earth somewhere between 25,000 and 160,000 miles per hour, it would hit our atmosphere and start squeezing the air itself. Compressed air heats up. When you pump up your bike tire you might not notice, but the air inside the tire becomes just a tiny bit hotter.* The meteorite is doing the same thing, only it's compressing a lot of air and it's doing it quickly.

Because of the compressed air below it, the meteor would become your own personal sun. The air around you would go from a cool 70 degrees to a scorching 3,000 in a matter of seconds. In that heat you would steam and blacken but probably not have time to ignite.

If you were left in a 3,000-degree oven the heat would eventually turn you into an expanding gas, but, mercifully, you would spend only tens of seconds in that heat before the meteorite thudded on top of you, so at least there would be something of you left, even if it were only a lump of coal.

It is not all bad news, though. You would have the distinction of being the first person to die by meteorite. However,

---

*Modified bike pumps, called fire pistons, can compress air so that it gets hot enough to start campfires.

you would not be the first person hit by one. That honor, as far as we can tell, belongs to Ann Hodges of Alabama, who was sitting on her couch in 1954 when a melon-size meteorite crashed through her roof, destroyed her radio, and hit her in the hip, resulting in a sizable bruise.

The second confirmed meteorite victim was Michelle Knapp's 1980 cherry-red Chevy Malibu. In 1992, Michelle heard a loud commotion in her garage, rushed out to investigate, and discovered her newly purchased $300 Malibu destroyed by a 26-pound, 4½-billion-year-old space rock.*

Fortunately for Michelle, Ann, and the rest of humanity, these meteorites were relatively small. It takes a rock the size of a fist to even make it to Earth intact—everything smaller gets burned up in the atmosphere—and fist-size rocks carry so little momentum that the atmosphere slows them to roughly 100 miles per hour. If a fist-size meteorite landed near you it would only be good news—meteorites can go for $100 per ounce.†

The largest meteoroid to hit Earth in modern times was the Tunguska strike that hit Russia in 1908. That rock was estimated to be 100 yards wide and hit with 300 times the power of the Hiroshima bomb. It made the loudest sound in recorded history and was deafening 40 miles away. It happened to land in northern Siberia and no one was killed, although 80 million trees were blown down by the shock wave,

---

*What started as a bad day for Michelle quickly began looking up, though, after she was able to sell the destroyed Malibu for $10,000 and the meteorite for $69,000.
†If you were lucky, it would be a meteorite from the moon or Mars. Those go for hundreds of dollars per carat, while the cheaper and more common ones come from the asteroid belt and fetch far less.

and a farmer who was 40 miles from the impact was thrown through the air by the blast.

Even if you weren't standing under it, a mile-size rock would be significantly worse news. If it entered the atmosphere at a low angle, the heat would incinerate everything below it as it passed overhead, leaving a clear path of scorched Earth.

Next would be the shock wave. A mile-wide rock would probably break up as it burned through the atmosphere, but the pieces would still hit with the same combined energy— the equivalent of a 500,000-megaton bomb (the largest hydrogen bomb ever detonated was 50 megatons).

And if it landed in the ocean? The water would hardly slow the supersonic, incandescent rock before it struck the bottom. Then the waves would form. The first wave from a mile-wide meteor would be more than 1,000 feet high and travel at Mach 1.* And that would be a small one. Much bigger waves would follow, with the largest arriving a few minutes later, after the displaced water reverberated back into the crater.†

All that being said, a mile-wide meteor is enough for incredible destruction, but probably not enough to extinguish life on the planet. While the dust and smoke it would kick up would cool the globe and cause widespread crop failure and famine, most likely it wouldn't wipe out all human life.

Given the danger meteoroids represent, lots of resources

---

*Too fast to surf, unfortunately.

†It's hard to imagine how devastating the tsunami would be. Just 2,300 years ago a meteor only 500 feet wide landed in the Atlantic and washed away what is now New York City.

are dedicated to spotting them early, although there's still nothing we could do if we saw one coming. If we're lucky, we'll see any potential planet killer a year or two out. If we're unlucky and the meteoroid comes from an unexpected angle, we'll have no warning at all, which is something to keep in mind if you find yourself underneath a twinkling star that keeps getting brighter.

What Would Happen If...
You Lost Your Head?

If your brain were snatched out of your head, you would die. Doctors decide whether you're dead by measuring your brain's electrical signals, and you need to have a brain to have any signals. So without one, you're kaput. Not surprising.

What is surprising is how much of your brain you could lose and keep functioning. You're probably thinking your brain is crucial, but remember, that's your brain doing the thinking—not exactly an unbiased source.

If you're a chicken, not only is your brain unimportant, you could do without your entire head. How do we know? Look at Mike the Headless Chicken, born in 1945 in Fruita, Colorado.

On September 10, 1945, Mike the chicken was headed for the dinner plate. His owner, the farmer Lloyd Olsen, took him to the backyard and chopped off his head with an ax. Much to farmer Olsen's surprise, Mike shook off the injury and carried on exactly as before—pecking at the ground for food (or at least trying to). Mike toured the country for two years before finally choking to death (he had to be fed with an eye dropper). How did he survive the ax?

Doctors at the University of Utah determined the blade had indeed removed his head but left Mike's brain stem intact. The brain stem controls basic functions like heartbeat, breathing, sleeping, and eating, which, if you get down to it, is about all a chicken does. Mike's arteries clotted before he could bleed to death and he was free to go about his business.*

In chickens as well as humans, the brain stem plays a crucial role in life from moment to moment, because without it you wouldn't be able to breathe or control your heartbeat. Damage any other part of the brain, and the results are less certain. The brain is malleable and can transfer jobs to other undamaged regions. It's also split into left and right hemispheres, and if the damage is isolated to one side it can withstand a shocking amount, as we can see in the case of Phineas Gage.

Railroad construction had somewhat lax safety standards in the early 1800s, particularly for the dynamite crew. Phineas Gage's job, as a part of that crew, was to pour gunpowder into holes bored in rock and then tamp it down with a 1¼-inch-thick, 3½-foot-long metal rod—but before striking the gunpowder

---

*If *your* head was chopped off, experiments in rats show that you would have around four seconds of consciousness before the massive loss in blood pressure, akin to getting up too quickly out of a hot tub, would cause you to pass out.

he had to make sure to add a bit of sand so that he wouldn't ignite it.

On September 3, 1848, Phineas Gage forgot to add the sand.

When he hit the gunpowder it exploded and fired the metal rod through his jaw, behind his left eye, through the left hemisphere of his brain, and out the top of his head before landing a few hundred yards away.

Not only did the bar not kill him, Gage never even lost consciousness. After a month he had almost entirely recovered, although according to his friends his personality did seem to change. The consensus, post-bar-through-head, was that he was more irritable. After the accident Gage left the railroad and went on a publicity tour with his bar and lived for another twelve years.

Gage was, of course, lucky. Though the bar passed through his brain, the damage was contained to the left hemisphere, and because some of the most critical functions have backups in the opposing side, if you are going to shoot a rod through your head it's much better to go front to back or top to bottom and destroy only one hemisphere than it is to fire the rod ear to ear and destroy both.

Another reason Phineas survived is that large chunks of the brain don't seem to be doing much of anything, or at the very least are redundant. If the damage happens slowly, you can afford to lose even more than Phineas did, as in the case of a student of the British neurologist John Lorber.

In the late 1970s, Lorber was a professor at Sheffield University in England and noticed that one of his honor students had a remarkably large head. He recommended the student get a CAT scan. The scan didn't just reveal a problem with the student's brain, it showed he barely had one at all—95 percent of it was cerebro-spinal fluid, with only a thin crust of gray matter pinned against his skull.

This condition isn't totally remarkable—it's called hydrocephalus, and it's basically like having a leaky pipe in your brain. The leaking fluid gradually pushes your brain outward against your skull. If it happens when you're young and your bones are still malleable, the pressure pushes out your skull as well—hence the large hat size.

What was remarkable about this student was that he had an IQ of 126 (a score of 100 is average), which might tell you something about the IQ test, but it also means that when it comes to brains, size doesn't matter all that much.* We have three pounds of brain stuffed into our heads while he was working with a quarter pounder and doing just fine.

For a while scientists believed the bigger the brain the smarter the animal (and that we had the biggest). Then someone took a look inside an elephant's skull and saw its twelve-pound brain, and the theory had to be amended. Perhaps it was the brain-size-to-body-weight ratio that dictated intelligence? That sounded good until someone did the math and realized it puts us on par with the field mouse.

In the end, the key to intelligence is probably how many neurons there are in whatever size brain you do have, and

---

*The other explanation for this student's remarkable mental capability is that the inner part of the brain (which he was almost entirely missing), called white matter, is not quite as important as the outer bits, called gray matter. So if you're going to lose some brain, scoop some out of the middle.

judging an animal's intelligence by the size of its brain is like judging the speed of a computer by its size (and remember, the phone in your pocket is many, *many* times faster than the room-size computers of the sixties).*

Basically, if we're ever invaded by pea-brained aliens—*do not* underestimate them.

---

*Incidentally, while we're on the topic of brains versus computers, yours can still do some things faster than the fastest supercomputer. But the computer is catching up.

What Would Happen If . . .
You Put on the World's Loudest Headphones?

WHAT IF YOU put on the world's loudest pair of headphones and cranked the volume to eleven? Would the death metal rattle your skull and liquefy your brain?

Fortunately, the answer is no. If you put on a pair of 190-decibel headphones, your eardrums would instantly rupture and you would be permanently deaf, but your brain can withstand more energy than music can deliver.

However, the same does not go for some of your other organs. Headphones keep the sound focused on your head where everything but your eardrums is resistant to acoustic energy, but if you unplug the earbuds and listen to speakers you will expose your whole body—and your eardrums are not the only cavity that's a little vulnerable to sound waves.

Before we get to that, though, it's important to understand what's going on when you're listening to music. Sound is a series of pressure waves moving through the air. You interpret those pressure waves as music because of wiggling bones in your ear that set off a Rube Goldberg–like system between eardrums, membranes, "hairs," bones, and electrical nerves.

A higher pressure wave of sound equals more wiggling and a louder noise. So sound is actually pressure waves moving through the air, which is why it can cause damage.* The most dangerous sounds are caused by shock waves, which result from major events like bomb blasts where the pressure goes from one atmosphere to many atmospheres in one or more pulses. While these are sounds, they don't qualify as music because the pressure wave is a single spike, whereas music is an oscillation of pressure. Because the loudest possible oscillation is between 0 and 2 atmospheres; the maximum decibels music can reach is 194. Anything louder would be a shock wave. Therefore the question "Can you be killed by music?" can be translated to "Can you be killed by a sound of less than 195 decibels?" Which begs the question, what is a decibel?

Decibels are used to measure volume and they are logarithmic, which means an increase in 10 decibels equates to a sound with 10 times more energy.

At 120 decibels—equivalent to standing next to a chainsaw—sound begins to get painful.

At 150 decibels you would feel as if you were standing next to a jet engine. The sound would resonate so intensely in your inner ear, it would blow out your eardrum. That would solve

---

*These pressure waves dissipate in the air as heat, and though yelling doesn't produce enough heat to be a health risk, if you hollered at a cold cup of coffee that was in a perfect thermos, your cup would be hot and ready to drink in a year and a half.

the too-loud problem, but if the decibels increase it could still do more damage.

If you released 190 decibels of sound out of speakers you could be in trouble.* Luckily that's not a practical concern. The loudest man-made speaker is a horn in the Netherlands used to test whether satellites can withstand the noise of a missile launch. The horn produces 154 decibels, which is enough to burst your eardrums but probably not enough to kill you unless you stuck your head in it for a while (scientists aren't sure because nobody has tried this yet).

Of course, the satellite horn is only the loudest one that we know of.

The U.S. military has experimented with sonic weapons since the 1940s but as far as we know has been frustrated by the results. In concept, the ears make for an inviting target. You can't close them, turn away, or refuse to pay attention. But in practice sound is difficult to control. It bounces off objects, can be amplified by buildings, and is ineffective for crowd control, where those near the speaker could be instantly deafened while people in the back would barely be annoyed. Perhaps worst of all for the military, a sonic weapon can be countered by a five-dollar set of earplugs.

But let's say you're attending a death metal concert, where they have cranked up the speakers to 190 decibels and you have a front-row seat. The sound would immediately blow out your eardrums and leave you permanently deaf, so the noise would not be heard so much as felt.

---

*One way to make a speaker this loud is to connect one exit tube alternately between one vacuum chamber and one chamber pressurized to two atmospheres.

Sound waves actually compress air as they pass through the atmosphere, but since your body is mostly liquid, it's almost immune to this compression. We say "almost" because not all of you is liquid. There are some hollow parts, such as your lungs and GI tract, and it's those hollow spots that you need to be concerned about.

Your intestines, luckily, are tough. It would take more than two atmospheres of pressure to rupture them. To break those open would require the shock wave from an explosion. Your lungs, unfortunately, are far more delicate.

Lung tissue is relatively fragile, and extreme sound vibrations could cause a rapid overexpansion and destroy the small alveoli sacs that line your lungs. Alveoli are the key intermediary between your lungs and blood that allow the gas exchange to take place. Without your alveoli, you couldn't oxygenate your blood and your lungs would be useless.

So if you were standing in front of a speaker listening to death metal and it was turned up to eleven—in this case, 190 decibels—the pressure wave would force your lungs to overexpand and perhaps break your alveoli sacs. You would suffocate while trying to breathe like a fish out of water.

Of course a true metal fan should travel to Venus. In our atmosphere 194 decibels is the upper limit for music, but on the surface of Venus, where the atmosphere is much denser, rock music can be ten thousand times more powerful. Listening to a guitar solo would be like standing near a bomb blast.

What Would Happen If . . .
You Stowed Away
on the Next Moon Mission?

Nasa is probably not going to return to the moon in the near future. In preparation for a manned trip to Mars, the current plan is to land on an asteroid instead. If you want to go to the moon, your best bet is get a ride with the Chinese. But even if you do speak Mandarin, the competition for the job is intense. Let's go out on a limb and say you won't be accepted. But what if you're determined? What if you refuse to take no for an answer, so you stow away on the spacecraft? And because space suits are expensive ($12 million) you just wear shorts and a T-shirt. Here's what we think would happen.

At the count of *wǔ* (five), which you wouldn't hear over the radio like the real astronauts but could probably hear from the loudspeaker outside, the main engine would fire. At liftoff the spacecraft would accelerate over the next eight minutes to 25,000 miles per hour and you would endure 4 g's of acceleration, about the same as the most intense roller coasters but over a much longer period. This is survivable, but without the G suits and padded seats that the astronauts are using you wouldn't be comfortable and would probably pass out. The space suits are also helpful if there's a breach in the spacecraft. Since you're not wearing one, you would need to hope for smooth sailing.

You would also need to hope the space agency added some extra fuel for the trip, because with your extra two hundred pounds of body weight, the spacecraft's trajectory would be incorrect and the engineers would have to fire maneuvering rockets to adjust your course.

But let's say all goes well, and by the time you're discovered it's too late to do anything but take you along. How would you be feeling traveling in zero-g for three days toward the moon? Very, very sick.

Nausea is an unfortunate part of the initiation into life in zero-g. Space sickness is a supercharged version of motion sickness, which is itself the uncomfortable result of a "disagreement" between your eyes and your inner ear. Your brain interprets this disagreement as food poisoning and prescribes an antidote: vomit.

Exactly how sick you would become depends on the quality of the connection between your brain and inner ear. Nobody's connection is flawless—if you are spun under water, your inner ear can't tell which way is up—but the higher your fidelity, the stronger the disagreement, and the sicker you will get.

The current space-sick champion is former Utah senator Jake Garn, who used his position on the Senate Appropriations Committee to earn himself a ride to space in 1985. The degree of Senator Garn's nausea became so legendary that NASA named the space-sick scale in his honor. The Garn Scale goes from 0 to 1.

At 0 Garns you're feeling fine, and your typical car-sickness nausea registers at only a tenth of a Garn. A full Garn means you're totally sick and incapacitated.

Vomiting is not typically lethal on a windy car ride, but in space it's dangerous. If you're on a space walk with a helmet on you can drown in your own sick.* So in order to help with the problem, NASA trains astronauts in a specially outfitted aircraft, nicknamed the Vomit Comet, which carries passengers on enormous parabolic loop flights. At the beginning of each parabola (just as it starts to climb), the plane begins to freefall, and for roughly ninety seconds everybody inside falls with it—zero-g.

In your case, lacking any Vomit Comet training, your inner ear would be thrown for a major loop, and very quickly you would hit a full Garn—near incapacitating nausea.

The good news is that once you landed on the moon, its gravity would cure you of your space sickness. The bad news is that you still wouldn't have a space suit.

The moon, just like space, is an airless vacuum, which is why your fellow astronauts will be wearing their expensive and cumbersome space suits when they step onto it. When you venture out onto the moon in your more comfortable outfit, you will die. But not instantly!

---

*Liquid of any kind in your space helmet is dangerous. In 2013, on a space walk around the International Space Station, Italian astronaut Luca Parmitano nearly drowned when water leaked into his helmet, sending dangerous globs floating about.

How do we know?

In 1966, a NASA technician proved it. While testing a space suit in a vacuum chamber, a faulty hose caused the suit to depressurize. He was in the vacuum unprotected for eighty-seven seconds before the chamber could be repressurized. For most of that time—all but the first ten seconds—he was unconscious. But fortunately, other than an earache from the rapid pressure changes he was unharmed. The lesson? In a vacuum a human body can survive for a minute—maybe even two—without protection but can only stay conscious for about 10 seconds.

What would you experience in that brief window of consciousness?

It depends on what side of the moon you are on. Are you on the sunny side, or the shady side? It makes a difference. Earth takes twenty-four hours to do a full rotation, but the moon takes an entire month, which means one side is allowed to bake in the sun for fifteen days and heats up to 253 degrees, while the shady side gets down to 243 degrees below zero. This temperature difference would matter when you first opened your door and stepped outside. What would you feel?

If the sun was down and it was 243 below, you would feel a chill but not freeze, because 243 degrees below zero in a vacuum is different from walking into a 243-degrees-below-zero freezer on Earth. Without any atmosphere, heat transfer happens slowly. If you landed on the shady side, the temperature change would feel roughly like stepping into a cool room naked. Then, because the boiling temperature of water is lower than your body temperature in a vacuum, you would feel a chill as your sweat instantly boiled off. But that's the worst it would feel—just a chill.

If you landed on the sunny side of the moon, where it's 253 degrees above zero, the vacuum would again save you from baking. But because of the radiating heat from the hot surface of the moon, you would feel just a bit warmer than on a summer day in Death Valley.

In addition to being a little warmer, there are a few other differences from the shady side. The moon surface is also 253 degrees, so without a boot on you would need to be careful where you stepped. Most of the surface is a fine powder, which isn't very dense. It's so light, in fact, that instead of the moon burning your foot, your foot would cool down the moon.* Step on a moon rock, though (they're everywhere and denser than your foot), and your foot would sizzle.

Along with avoiding moon rocks, you would also need to take into consideration the sun—more specifically, its UV radiation.

The sun is firing X-rays, ultraviolet light, and high-energy particles of radiation at all of us all the time. Fortunately for everyone on Earth's surface, the planet's atmosphere, ozone layer, and magnetic field take care of most of that, and sunscreen or clothing blocks the rest.† Under these layers of protection, life can thrive. For anyone above the atmosphere, however, the situation is radically different.

On the moon you would not have the benefit of the atmosphere's protection, so even if you carefully applied SPF 50

---

*This is the same reason why walking on hot coals is possible if done correctly.
†Earth's atmosphere has an SPF rating of around 200.

before stepping out, in a few seconds you would get enough radiation to give you a healthy tan. Within fifteen seconds you would absorb a dose that would eventually develop into a blistering third-degree sunburn.

Another consideration is breathing. If you took a deep breath and held it before you left the lunar module, the air in your full lungs would instantly expand in the vacuum, ripping apart the delicate alveoli sacs. The best way to deal with this is prevention: Instead of filling your lungs or holding your breath as you exit the craft, you would need to keep your mouth open, letting the gas in your lungs rush out.

Your blood contains enough oxygen to give you ten to fifteen seconds of consciousness. After that you would pass out, and 1960s research into vacuum exposure with dogs shows that after two minutes you would be brain-dead.*

Once your heart stopped beating things would get gruesome.

We said earlier that the boiling temperature of water is below body temperature in a vacuum, so all your sweat would boil off (along with your tears and saliva, which makes a stinging sensation), and that's true. But that's the water on the *outside* of your body. The water inside you—namely, your blood—would take tens of seconds to start to boil.

You would be unconscious and soon dead—so this is more

---

*Okay, we know this all sounds like bad news, but the opportunity for rescue is real! The experiments with dogs showed that exposure to vacuum for ninety seconds is nearly always survivable, though during this period the dogs were unconscious and paralyzed, and the gas escaping their bowels caused defecation, vomiting, and urination. Their tongues were also coated in ice and the dogs swelled up, so it doesn't sound pleasant, but after repressurization they deflated, and after a few minutes they were as good as new. Two minutes of vacuum exposure appears to be the limit, though.

of a cosmetic issue than anything else—but as your blood boiled and turned into a gas, your skin would expand until stretched taut, fully inflating you into a human balloon.

Eventually all that gas would escape your body and you would deflate, but the process of ripping your skin from its anchors would probably result in at least a few new wrinkles.

There aren't any bugs or bacteria living on the moon, just the ones living inside you, but they would be killed by the vacuum and wild temperature swings as well, so you would not rot or decompose.

Assuming your fellow astronauts didn't want to haul you back, you would stay on the moon for many thousands of years as a well-preserved, desiccated, and wrinkly moon man.

What Would Happen If . . .
You Were Strapped into Dr. Frankenstein's Machine?

A CLOSE EXAMINATION of the original Frankenstein texts doesn't reveal exactly what voltage or current the doctor used for his machine, but it would have to have been substantial. In any event, let's say you stepped in for the monster and strapped yourself to the table. Because you are, presumably, alive, not dead like Frankenstein's monster preelectricity, the current would have a far different effect on you than it would on the monster (and in reality be far more effective at making you dead rather than the other way around).

The first thing Dr. Frankenstein would do is strap electrodes to your head and ankles to guide the electricity through your body. Then he would throw the switch and a number of things would happen very quickly—but before we get into the specifics of what they would be, let's pause for a moment to talk about the electricity in your body right now.

A sharp electrical jolt is hitting your heart as you read these lines. At least you had better hope it is. If not, you are experiencing what doctors call *being dead.* Hopefully, if all is going well, your heart will be jolted eighty-five thousand times today, just like yesterday and, if you're going to have a tomorrow, just like tomorrow too.

The timing and the amount of electricity hitting your heart is critical, and it's easy to screw up. Your heart needs only a tenth of a volt to trigger its contraction, and a mistimed volt can mess up the beating of your heart and kill you.

That's the bad news.

The good news is your skin makes for a decent electricity-resistant suit.

If you're hopping onto Dr. Frankenstein's table and you're dry and wearing clothes, anything less than 100 volts would probably not make it to your heart.*

To guarantee the current's access to your innards, Dr. Frankenstein would need to use at least 600 volts—powerful enough to force a dielectric breakdown or, more colloquially, blow a hole in your skin.

Next, your body would jump as the electricity mimicked the voltage your nerves use to fire your muscles.† This jolting

---

*It's difficult to say exactly what the lethal dose of electricity is because how it travels is a little unpredictable. At least one person has been killed by as few as 24 volts, but that involved a lot of water.

†Part of the danger in grabbing an electric fence is that the electricity will cause your arm muscles to fire—and the clench muscle is stronger than the release muscle, so you cannot let go of the fence. The same goes for your legs. Electricity does not "blow" people off the ground. It causes leg muscles to fire, and the extend leg muscles are stronger than the contract ones, so you jump.

phenomenon is the origin of the Frankenstein story after author Mary Shelley saw corpses jump in experiments, which apparently sparked an idea ("It's ALIVE!!").

Mild electrical stimulation is not necessarily bad. Electrical shocks that force muscles to contract over and over is called exercise—six-pack abs without effort!

You would have other issues beyond unwanted exercise, though. The current would not want to travel along your skin where the resistance is high, so it would worm its way into your brain through the low-resistance pathways of your nose, eyes, and mouth. Whatever the current touches it heats, which is not so bad on your skin—just a light scorching and smoldering. However, your brain is more sensitive.

Once the current gained access inside your skull it would heat and cook your brain's proteins. After scorching the outside of your brain, the current would continue on its way toward the electric band on your ankles, which means it would pass and concentrate on your brain stem, where a number of vital functions are controlled, like breathing. Once the brain stem is fried, you would "forget" to breathe, no matter how hard you tried to remember.

Your brain can continue to function for a few seconds on oxygen reserve, but within fifteen seconds it would go unconscious, and after four to eight minutes you would experience complete brain death. If this were Mary Shelley's story, brain death would of course be no problem. Dr. Frankenstein could just throw the switch again and you would be up and walking in no time. In reality, however, brain death is a more problematic condition. If your heart is fluttering it can be reorganized by a jolt of electricity, but trying to jump-start your brain is like trying to jump-start a computer.

Not to mention your brain has already been denatured, so if Dr. Frankenstein wants to reverse the process and bring you back to life, he will have to start by digging up a new brain.

What Would Happen If . . .
Your Elevator Cable Broke?

OVER THE 150 years of modern elevator history, on more than 800 billion rides, it's likely a majority of the 1.3 trillion elevator passengers have at one time worried that the cable would accidentally break and they would die a horrible, flattened death.

And they have reason to.

Because it has happened.

Once.

In 1945, the pilot of a U.S. Air Force B-25 became lost in the fog and flew into the seventy-ninth floor of the Empire State Building, severing the hoist and safety cables of two elevators and sending both plummeting down the shafts. In those days, before elevators became automatic, they had operators—people who sat inside the elevators and guided passengers to their destinations.

One of the operators had stepped away on one of the best-timed smoke breaks in human history. The other, Mrs. Betty

Lou Oliver, plunged seventy-five stories into the elevator pit below.

Elevators are the safest motorized transport you can use. They are not entirely without risk—on average, twenty-seven people die in elevator accidents in the United States every year, but nearly all of them are due to "operator error." That would be you. (Safety tips: Don't squeeze your way into closing elevator doors. Don't try to climb out of a stuck elevator. And don't ride on top of them.) In comparison, escalators are thirteen times more dangerous.

Part of the reason elevators are so safe is thanks to the safety brake, invented by Elisha Graves Otis in 1853. The safety brake is on the elevator car itself and allows an elevator to stop even if the cable is severed.

Elevators weren't popular until Otis's invention. Before then, nobody wanted to get into a box where their life hung by a single thread, even if it was a thick one. Otis changed that, and when he did, he changed everything.

Elevators might seem like no more than a handy-dandy modern convenience, but, in fact, the elevator is essential to urban living as we know it. Before elevators, buildings topped out at six stories—nobody was willing to carry up a bag of groceries any farther—and in pre-elevator buildings the penthouses were on the first floor. The fewer stairs you had to climb, the more you paid for your apartment.

Elevators allowed architects to build up, increasing the number of people who could be jammed into a city block. Without elevators, our population would have oozed outward from city centers in a never-ending suburban sprawl.

Thanks to Mr. Otis, not every city looks like Los Angeles, but should the impossible happen, should Otis's invention fail, and your elevator plummeted from the top of a skyscraper like Mrs. Oliver's, you wouldn't necessarily die. With a bit of luck, and because of a few freaks of physics, you could survive—just like she did.

The farthest you could possibly fall in an elevator these days is 1,700 feet. Elevators can't go higher because their hoist cables become too heavy; it wasn't until the invention of an elevator-transfer floor in the World Trade Center in 1973 that skyscrapers exceeded this elevator limit.

An elevator free-falling from 170 stories would hit the ground at 190 miles per hour—an almost certainly fatal speed. But if you're lucky, your elevator fits snugly in its shaft. If that's the case, the air below won't be able to escape fast enough, creating a pillow of pressure like a soft airbag that could slow your descent.

That would help, but you would still need to do more to survive.

Gradually slowing your stop is the key to reducing the g-forces on your body. G-forces are a way to express the force of acceleration or deceleration on your body using Earth's gravity as a unit of measure. Right now you're experiencing 1 g. The most intense roller coasters peak around 5 g's (which means you will "weigh" 5 times your weight). Trained fighter pilots can withstand 9 g's and keep flying.

Around 50 g's over a few seconds appears to be the surviv-able limit. How do we know? In 1954, the U.S. Air Force was designing fighter jet ejection seats and needed to know how fast they could get pilots out of their planes without killing them. Specifically, they needed to know how many g-forces the human body could withstand. So they built the world's scariest carnival ride and asked for volunteers.

Air force officer John Stapp, already having almost suffocated himself testing oxygen systems and nearly skinning himself by flying without a canopy at 570 miles per hour, got the call.

The air force strapped Mr. Stapp into a specially designed rocket sled, accelerated the sled to Mach 0.9, and then watched what happened when it stopped in just 1.4 seconds, equating to 46.2 g's.

For one very uncomfortable moment Stapp "weighed" more than 4,600 pounds. The blood vessels in his eyes burst, his ribs cracked, and he broke both wrists. But he survived, and proved that—properly restrained—you could withstand more than 40 g's of deceleration.

One of the reasons John Stapp survived was his positioning, which takes us back to your free-falling elevator. Your best chance is to load your entire body evenly. Do not jump. It does not help. Even if you somehow, magically, jumped an instant before you hit, you would reduce your impact speed only by a mile or two per hour, and when you crashed, your organs would break from their arterial moorings and push their way down through your body.

You shouldn't hang on the ceiling light fixture, in case you were thinking about that. You would be ripped off and then slammed into the floor just as hard as if you had jumped from the top floor. And as tempting as it might be, climbing onto the shoulders of the person next to you would not help either. It's precarious, and he or she would just topple over at impact anyway.

Best practice? Lie on your back. It's the best way to bring your body to a stop without causing an organ pileup.

Interestingly enough, when Mrs. Oliver was discovered inside her shattered elevator she was not lying flat on the ground, as we have recommended—she was sitting in the corner. Amazingly, she survived even though sitting isn't the perfect position. She suffered broken ribs and a broken back, but had she been lying flat on the ground she probably would have been speared by debris that was at the bottom of the shaft and pierced the bottom of the car.

So don't be misled. If your elevator cable breaks, your odds of surviving are pretty slim. Fortunately, the odds of it happening in the first place—less than one chance in a billion—are much slimmer.*

---

*As a comparison, if you're trying to get to the second story of your building, taking the stairs or escalator are both ten times as dangerous and rock climbing the outside of the building is more than a thousand times riskier.

What Would Happen If . . .
You Barreled over Niagara Falls?

In 1903, SIXTY-THREE-YEAR-OLD retired schoolteacher Annie Edson Taylor was strapped for cash. Staring down a future in the poorhouse, she decided to be the first person to go over Niagara Falls in a barrel, thinking it would lead to fame and riches.* She constructed a barrel, pressurized it with a bicycle pump, and sent it over the falls with her cat inside as a test. The cat and barrel survived, so on her birthday she had herself towed out into the middle of the river and dropped off. A few minutes later her barrel was retrieved at the bottom, along with a relatively unscathed Ms. Taylor. In spite of her success, she is quoted as saying "I would sooner walk up to the mouth of a cannon, knowing it was going to blow me to pieces, than make another trip over the Falls."

Despite this advice, Ms. Taylor's survival inspired many

---

*It didn't.

copycats, many of whom were not as lucky. Barrels are the most popular craft, but other vessels have included a kayak, a Jet Ski, and even a giant rubber ball.

Let's say you, like Ms. Taylor, chose a barrel as your craft, had a friend drop you into the middle of the Niagara, and allowed the current to drag you over the falls.

By the time you reached the bottom you would have fallen 180 feet and sped up to 70 miles per hour. Whether you survived would depend on what you hit.

If your barrel hit the rocks you would be in trouble. In NASA studies testing the human body's durability, they concluded that an unrestrained fall of 22 feet—which means you would hit at 25 miles per hour—onto something solid and landing on your feet is usually survivable (this does not mean you wouldn't suffer catastrophic injury—you probably would). A fall from 23 to 40 feet means your survival is questionable, and falling 40 feet (at 34 miles per hour) onto rocks is almost certain death.

Clearly, if you and your barrel hit rocks traveling at 70 miles per hour at the bottom of the 180-foot falls, you would die.

Landing in the water below the falls is far more preferable than landing on rocks, so your best chances are in the Horseshoe section of the falls, where you would fall into water. However, that doesn't mean you would be safe, especially if it's a pool of still water. Studies conducted by the U.S. Air Force show that if you hit still water at 70 miles per hour, your survival chances are just 25 percent, and that's if you hit the water perfectly (feet first, knees a little bent, body leaning slightly back). Hit it in any other position and it's near certain death.* That's because if you did not hit perfectly you would do nearly all your slowing down in the first foot of water and the delicate bones of your rib cage would shatter under the intense g-load, firing pointy spears at your organs. Your skull would fracture as your head crushed into your spine. That goes for your other organs as well, which would carry their momentum toward your feet.†

There is some good news, though. The water below Niagara Falls isn't still. It's aerated, agitated, and churned up, which is good for high-speed landings. Air bubbles are less dense than water, so you would travel farther into the frothy water before stopping, decreasing the g's you would experience. The aerated water below Niagara is what allows so many stuntmen to emerge with their organs still intact.

The bad news is *also* that the water is aerated and churned up, because that makes it less dense, which means you don't float in it. This is probably why, no matter how unseaworthy barrels may seem, people going over the falls in sealed barrels actually do survive more frequently than those who go over in their swim trunks. Barrels float better than people.

If you survived the fall with barrel and body intact, the

---

*According to the same study, a 240-foot fall (hitting the water at 80 miles per hour) is fatal regardless of body position. The Golden Gate Bridge is 245 feet high, and 95 percent of jumpers die on impact.
†Common question: If you were falling into water, would firing a bullet into the water to "break up the surface tension" as you fell save you? Unfortunately, no. Surface tension is not relevant to your survival. What is relevant is the density of the water and how quickly it will stop you. In order to survive you have to decrease that density, so what you need are lots of bubbles, and a single standard bullet will not create enough. To survive you would need a column of bubbles three feet deep and as wide as you are. So to give yourself a chance, you would need either an explosive bullet or many bullets, e.g., a machine gun.

next issue you would face is how the water recirculates under the falls. Barrels are sometimes stuck for hours behind the curtain of water.

George L. Stathakis, another Niagara daredevil, who went over the falls in 1930, was stuck in his barrel for fourteen hours behind the curtain. Even if your barrel maintained its integrity, it would not have enough air for you to last that long. Sometime during his recycling under the falls, Stathakis suffocated.

The recycling currents under Niagara are the real killer. The aerated water usually saves daredevils from death on impact (though many break bones), but how it recycles you is random. If you were lucky the current would spit you out within seconds and you could go on a publicity tour to pay off the fine you would be assessed. If you were unlucky, like Stathakis, you would be pulled under, stuck behind the curtain, and buried alive in water.

What Would Happen If . . .
You Couldn't Fall Asleep?

On your 10,000th day alive, you will have spent 27 years, 4 months, and 25 days on this planet. Or, if you prefer, you will be 240,000 hours old. Of those hours, you will have spent 11,000 eating, an entire year in the bathroom, and another year with your eyes blinked closed. All of that is dwarfed, though, by one of your favorite activities—being unconscious. By the time you're 10,000 days old, you will have spent 9 years asleep.

If you were given the chance to get all that time back, would you take it? Or, to put it another way, would you drink the *ultimate* energy drink, one that enables you to stay awake *forever*?

Think carefully before you answer. If you're faced with a choice between going foodless or going sleepless, you should forgo the ham sandwiches. Going without sleep will kill you more quickly, and in much more distress, than skipping food.

The more interesting question is . . . *why?* Experts aren't totally sure. Whatever is going on during sleep, it's obviously important, not just because of the enormous amount of time devoted to it but because evolutionarily it doesn't seem to make sense. For much of our history we shared a world with some very large predators. We occupied no more than a middle rung on the food chain. Lying down completely oblivious to an approaching saber-toothed tiger for hours at a time seems hazardous. It's hard to imagine, in an environment where only the fittest survived, that the fittest included animals that were sitting ducks for a third of their lives.

Clearly something important is going on here. Sleep is almost a universal need throughout the entire animal kingdom, whatever the risk. Mice will doze in an environment filled with cats. Even plants have a daily circadian rhythm that's sleeplike.

Obviously sleep is an adaptation that goes far back in evolutionary time. Perhaps your distant relative—maybe some algae a few millennia ago—caught a few z's that cleared its blue-green head and allowed it to perform a little better than its peers. The rest is evolutionary history.

Although we don't know the name of that algae, we do know Randy Gardner's, and Randy provides a more recent insight into the essential value of sleep.

In 1964, Randy Gardner, a sixteen-year-old high school sophomore from San Diego, California, performed the longest medically observed insomniac feat in history. Guinness no longer tracks records like Gardner's (too dangerous), but in 1964, under official continuous observation, the sopho-more didn't sleep for 264.4 hours. That's more than 11 days.

It was part of a high school science project—hopefully a large part—and it did not go smoothly. On the third day he mistook a street sign for a pedestrian, and by the fourth night he was convinced that he was a professional football player. According to his medical examiners, he took great offense at those questioning his skills.

On the sixth day he began to lose muscle control and short-term memory. When asked to count backward from one hundred by sevens he forgot what he was doing halfway through the task. On the last day, though, he was still able to beat one of his observers in pinball (one questions his oppo-nent's skill). Despite all that, after fourteen hours of sleep, Gardner made a full recovery.

While Randy Gardner didn't take sleeplessness to his physical limit, unfortunately for a few rats, we think we know what happens when you do.

Researchers once forced a group of lab rats into insomnia by monitoring their brain waves and spinning a wheel under their feet when they began to drift off, forcing them to move. In other words, they could not sleep. *Period.*

After two weeks of this, the rats were dead. The research-ers then repeated the experiment, only this time they tried to save the rats with something other than a nap. During the experiment, the rats' body temperatures began to drop, so the experimenters raised the temperature of their environment. It didn't help. They saw the rats' immune systems weaken, so they fed them antibiotics. This didn't do anything either. The rats lost weight, so they were given more food. They still died. The only thing the researchers could do to save the rats was very simple: Allow them to sleep. After that, their recovery was nearly always complete. In some badly understood way, insomnia was "poisoning" the rats and the only effective antidote was sleep.

In people we can see the effects of insomnia by measuring brain waves. When you're tired, your prefrontal cortex, the part of the brain that controls memory and reasoning, goes into overdrive. It has to work harder to do the same amount of work it can do easily when you're feeling fresh, like an old computer opening a large file. Your brain just doesn't work well when it's tired.

To this day the only 100 percent definitive reason scientists can offer for sleep's necessity is, as Stanford University sleep researcher Dr. William Dement told *National Geographic* without trying to be funny, "we sleep because we get sleepy."

But that may be changing. Recent research may shine a bit more light on the subject.

In observations of both mice and monkeys (though not yet humans), sleep looks like it might be a kind of brain dishwasher.

When you're awake, your brain cells produce toxic waste proteins that hang around and impair brain function.* To clean these toxins out you have cerebrospinal fluid that washes through your brain cells and carries away the waste. Unfortunately, cerebrospinal fluid doesn't flow when you're

---

*One of the waste products removed during sleep is called beta-amyloid, and its presence is closely associated with Alzheimer's and dementia.

awake. When you're up and about, your brain cells are fatter so there's not much room to move between them. That means the spinal fluid gets "stuck" in a cerebral-fluid traffic jam and the toxins stay in place and build up.

Once you fall asleep, your brain cells shrink and your spinal fluid kicks into gear like it's midnight on the freeway. The fluid rushes through your brain and carries away the polluting toxins. When you wake your cells are fresh, clean, and ready to ponder life's deepest meaning, or whether you're going to have eggs or cereal.

If this theory is true, it explains why mental function falls so sharply when you're tired, why sleeplessness will eventually kill you, and why rats stubbornly refuse to stay alive when subjected to forced insomnia. Just by being awake you're dirtying your brain, and the brain, it appears, really, *really* hates being dirty. It desperately wants to sleep, as you might have experienced while trying to pull an unsuccessful all-nighter. It's so desperate to sleep that while many people have died by refusing water, warmth, or food, no one in medical history has ever been able to deny sleep to the point of death.* The impulse appears to be ultimately irresistible.

It seems evolution gave you the ability to sleep and it made damn sure you were going to use it.

Nearly fifteen hundred people die every year in car accidents because a driver's brain sent itself into an unconscious state despite knowing full well it was in charge of a one-ton

---

*There is a very rare and fatal disease called fatal familial insomnia that prevents its sufferers from sleeping, but it looks like the damage it does to your brain is what kills you and insomnia is a side effect.

object moving at sixty miles per hour. And that's just the beginning. Train, plane, and industrial accidents all the way up to Chernobyl have been blamed on drowsiness. If you're driving a train or car, drowsiness is a dangerous period that can lead to a microsleep, which is a short thirty-second or less period of unconsciousness. Microsleeps are impossible to resist, and falling in and out of them is so seamless you probably wouldn't be aware it happened, unless of course you woke up in a ditch.

Sleep may be the only human need so strong that you could never die in want of it. It could be that the only way to truly test your brain's ability to go without it is to hook yourself up to a larger version of the diabolical machine the unlucky rats died on. We don't recommend it, but if you did, approximately two weeks after stepping onto this torture contraption, after hallucinating conversations, being unable to hold a thought for longer than a few minutes, and perhaps believing yourself to be a professional football player, you would die a filthy brain-cell death.

What Would Happen If . . .
You Were Struck by Lightning?

On April 2, 1978, a Vela spy satellite designed to spot the tell-tale double flash of a nuclear bomb picked up a hit. Somebody, it seemed, had dropped a nuclear bomb on the small mining community of Bell Island off the Newfoundland coast. That seemed unlikely to military analysts—Newfoundland was an unexpected place for the Cold War to turn hot—and, indeed, a few quick phone calls confirmed that the mining community was not a nuclear wasteland.

So what had happened?

The Vela satellites ignored lightning because the flash from a nuclear bomb is far brighter. What the Velas didn't account for were superbolts—the rare lightning strikes so powerful they mimic nuclear blasts. The Bell Island strike was a super-bolt. Heard more than 30 miles away, it left a 3-foot crater, damaged houses, and exploded TV sets.

What is a superbolt? Normal lightning strikes from the

bottom of a cloud, just 3,000 feet above the ground. A one-in-a-million superbolt strikes from the top—30,000 feet above Earth—and because it requires far more voltage to travel the greater distance, a superbolt is more than 100 times as powerful as regular lightning.*

Superbolts are extremely rare and most occur over water, so only a few firsthand accounts exist: On April 2, 1959, a strike in Leland, Illinois, left a 12-foot hole in a cornfield, and in 1838 a superbolt struck the 800-pound mast of the HMS *Rodney* and "instantly converted it to shavings," according to Frank Lane in *The Elements Rage.*

So what would happen if you were really, really unlucky and stood underneath a particularly ominous-looking thundercloud that started generating electricity at its top, 30,000 feet above the ground? Would you be converted to shavings?

Probably. But the exact answer depends on exactly how the bolt strikes you and how much energy it delivers. Even a normal bolt of lightning could turn you into the mast of the HMS *Rodney* if the entire arm-wide bolt passed through you. They usually don't, though, even in direct hits, because lightning ordinarily strikes victims with only a portion of its force. Some people have even survived direct strikes when the bolt "encased them" instead of passing directly through their bodies.

Becoming encased in lightning sounds like a fatal experi-

---

*How a cloud generates electricity is still not totally understood, but we think it has to do with how ice and water travel up and down in the up and down drafts of a storm cloud, generating bits of static electricity in the exchange, like lots of tiny wool socks rubbing on a carpet.

ence, but if you're going to be hit, it's your best chance at survival—and it helps if you're wet. Electricity always travels along the path of least resistance, so if the bolt hits you and you're really wet, that path might be along the outside of your skin and not through you. The strike will also charge the air immediately around you, and for an instant can turn that air into an easier path than the one through your gut.* This is called the flashover phenomenon, and some people who have been struck and knocked unconscious woke up naked after the water on their skin was instantly vaporized and blew the clothing off their bodies.

One of the biggest differences between a lightning strike and a typical household electrocution is how quickly the lightning passes through you—typically between eight to ten microseconds. In regular, run-of-the-mill, fork-in-a-socket electrocutions, the timing of the electrocution and your heartbeat isn't as critical because the electric flow lasts a long time. In lightning strikes, exactly when the lightning passes through your heart can save you or kill you. If you're unlucky, it would hit your heart a moment before it contracts. If the current passes through your heart during this instant—which lasts only one tenth of a second—it would likely send your heart into fibrillation, which, without a defibrillator, is certain death.

But even if you were lucky and the lightning came a moment after contraction, you would still be in danger. A superbolt can

---

*If you feel static electricity building, like the hair on your arms rising or the air around you beginning to crackle, take cover quickly. Getting into a car is best. The metal of a car provides the ultimate low-resistance path for the lightning bolt. The electric charge follows around the outside of the car, avoiding the inside entirely.

scramble an entire town's wiring, like it did to the houses on Bell Island. Imagine what it could do to yours. Your brain works on tenth-of-a-volt signals. A lightning strike can overstimulate your central nervous system, temporarily overwhelm your brain, send you into unconsciousness, and possibly scramble your brain stem, which is the area that reminds you to breathe. If it's sufficiently scrambled, you would forget to do that.* This can happen even if you're not directly hit.

How can you avoid all this unpleasantness? Standing underneath a tree in a storm is a particularly bad idea.† Lightning can hit the tree, travel to the ground, and turn the area around it into a hot plate of electricity. That's bad for you because you're mostly just salty water, and since salt water has a lower resistance than the water on the ground you would become the path of least resistance.

The bolt would travel up one leg and down the other, hijacking your electrical system and causing your leg muscles to fire, forcing you to leap into the air. The current would also puncture and destroy the walls of the cells it passed through in a process called electroporation that would create a highway of dead material perfect for the growth of infections. The upside? At least the current won't pass through your brain stem, so you would have a shot at remembering to breathe.

When the superbolt hit the mast of the HMS *Rodney* it

---

*This is why CPR is important after lightning strikes. The brain stem can unscramble itself and you can start breathing again, but it needs time and that's time you won't have unless someone helps you breathe.

†Another bad idea? Lying in a ditch. The electricity traveling along the ground will arc through your body and to the other side of the ditch. Standing in a shallow cave isn't great either because of the arcing problem. Find a car and get into it.

flash-boiled every last bit of water in the mast, rapidly expanding the water molecules into gas and exploding the mast into the sea "as if the carpenters had swept their shavings overboard," according to Lane.

If you took a direct shot from a superbolt, most of the electricity would, in all likelihood, pass alongside you. However, superbolts are powerful enough that even if most of the bolt passed alongside you, there would still be plenty left over to stop your heart and scramble your brain. In other words, you would be dead—you just wouldn't have exploded.

But if you're really unlucky—or let's say you make the ill-advised decision to hold a metal rod high above you—and you take the full, dinner-plate-size bolt to the head, you would end up like the mast of the HMS *Rodney*. The electricity would travel down your juicy veins and organs, heat you with more energy than if you were standing on the surface of the sun, turn your water to steam, and explode you into tiny bits.*

Vela satellites would pick up the strike, and perhaps a few scientists would fly in to make sure no one set off a nuclear bomb, but all they would discover would be a few broken TV sets, a couple of rattled neighbors, and one evenly distributed human.

---

*Even a regular lightning bolt can pass through your skin, heating and rupturing capillaries and creating etched patterns called "lightning flowers" or "skin feathering."

What Would Happen If . . .
You Took a Bath
in the World's Coldest Tub?

W<small>E HAVE ALL</small> accidentally run a bath that was too cold, but what if you really screwed up, things got way out of hand, and you took a bath in the world's coldest tub? Maybe the plumber somehow made a mistake and switched the cold water with liquid helium, the world's coldest liquid, and let's say instead of dipping a toe in first you hopped right in.

This actually almost happened to a few scientists. (Okay, not this *exact* scenario, but close.) Nine days after reopening the Large Hadron Collider—the giant particle accelerator in Switzerland—a solder joint failed and six tons of liquid helium spilled into the tunnel.* It was pure luck that no one happened to be there when it happened. If any scientists had

---

*The failure was spectacular. The electrical energy of a small city was accidentally dumped into the metal around the joint, instantly vaporizing it, and the resulting explosion moved a ten-ton magnet more than a meter.

been in the tunnel they would have been (spoiler alert) frozen like the bad guy in *Terminator 2*.

Helium is the gas you're probably familiar with from party balloons. It has to be 452 degrees below zero to be a liquid, which is just a few notches above absolute zero.

If your tub were filled with liquid helium, some of it would warm up and convert to gas, and one pound of liquid helium produces one hundred cubic feet of gas. That's going to displace quite a bit of oxygen.

So once you hopped in, your shriek would probably come out as a squeak. That's because the speed of sound is more than twice as fast in helium as in air, and the quality of your voice is determined by how sound reverberates in your mouth. In helium it bounces faster and your voice rises an octave.

So you would sound funny.

Of course, there's also the cold problem, but for at least a few seconds after you jumped in you might be surprised by how little pain you felt. That's because of something called the Leidenfrost effect. When you first hit the extremely cold liquid your warm skin would instantly convert the liquid helium it touched to a gas, which would insulate you from the extreme cold. The Leidenfrost effect is the same reason you can painlessly dip your hand into liquid helium, liquid nitrogen, or even molten lead if you're fast enough.

We're not sure exactly how long the effect would last, but you could plan on at least a few relatively pain-free seconds.

Eventually your skin would cool enough that it would stop

boiling the liquid and the helium would hit your skin. This is when the pain would begin.

You have two types of receptors primarily responsible for telling you you're cold. One tells you you're chilly—that one activates at temperatures down to 68 degrees—and another tells you you're freezing, a signal that you interpret as pain. That freezing neuron begins firing when you touch something below 60 degrees. The colder it is, the more pain you feel.

Needless to say, in liquid helium you would bypass the chilly neurons and go straight to extreme agony. But in addition to the pain, you would also be dealing with another issue: asphyxiation.

All the helium gas you would be breathing not only would make your yelps sound funny, it also would be displacing oxygen. Helium isn't poisonous, which is why you can inhale it from a balloon as a party trick. However, in this case, it would displace enough oxygen to turn deadly, and because your body can detect only a rising level of carbon dioxide in your blood and not a decreased level of oxygen, you wouldn't realize there was a problem. As soon as you got in the tub, you would have only fifteen seconds of consciousness before you passed out.*

Somewhere between your first high-pitched squeaks of pain and passing out from lack of oxygen, a window of probably ten or so seconds, you might notice something funny going on with the liquid.

---

*Why does your body recognize only CO_2 and not O_2? Detecting O_2 levels is a difficult bit of chemistry. But CO_2 in your blood raises its acidity and detecting acidity is easy, whether in your body or in chemistry class—so evolution probably just took the easy way out.

Liquid helium is known, of course, for being very, very cold, but it's also one of a select group of liquids called super-fluids because it seems to have a few superpowers.

For one, it has so little friction that if you stirred up a tank of it and came back a million years later, bits of it would still be swirling.* It can also climb walls. It is so light, and so frictionless, that if you poured it into a glass, it would crawl its way up the lip of the glass and drip onto your hand, which means that if the tub was filled to chest level, the liquid would climb all the way to your neck.

Your neck is a bad place for supercold fluids. It doesn't have much insulation and it's carrying a lot of blood. Even if you didn't pass out from lack of oxygen (say you brought a scuba tank with you), the liquid helium would freeze your blood to create ice dams within your neck. Your brain needs blood to work, and once your blood is blocked in your arteries, it wouldn't be getting any, so it would fail.

Even after you were dead, though, you would keep freezing. Soon you would be rock solid like the bad guy in *Terminator 2*—and, yes, if someone shot your frozen body with a bullet, some parts of you would shatter.

You do have a few advantages over the Terminator when it comes to supercold liquids. Because metal is a fantastic conductor of heat, even though it was only the Terminator's feet that were covered in the liquid (in his case, the slightly warmer liquid nitrogen), his entire body froze. Your flesh is a

---

*It's too bad about the whole freezing-to-death thing, because if it weren't for that, liquid helium's super-low friction would make it great for slip-and-slides.

much better insulator, so if you just put your feet in the tub, your head would not turn to ice.

However, you also have some disadvantages compared to the Terminator. Namely, once his body defrosts he's good to go. You would not be.

Eventually the helium would evaporate and you would thaw out, and the thawing out is what would kill your cells. This, incidentally, is the problem (okay, one of the problems) that the brains in cryogenic labs face. If you were to freeze slowly, the water in your cells would grow spikes like a snowflake and these spikes would destroy your cells. If you were frozen quickly, however, in a tub of liquid helium or in a cryogenic lab, you would skip the spiky snowflake stage and your cells wouldn't be permanently destroyed.

Unfortunately for you, and for all the heads in the cryo lab, there is no way to quickly go from frozen to unfrozen, so in the transition back to room temperature, your cells would grow those spikes and die.

Destroyed cells are dead cells, and dead cells can't be revived once they're gone—so, unlike the Terminator, you won't be back.

What Would Happen If...
You Skydived from Outer Space?

THE HIGHEST SKYDIVE in history was Alan Eustace's jump from 29½ miles above New Mexico in October 2014. He fell at 822 miles per hour, breaking the speed of sound and setting off a sonic boom that could be heard from the ground. Alan, however, did not begin his jump from space—the somewhat arbitrarily designated line 62 miles above Earth*— for a few good reasons. But let's say you didn't listen to reason and decided to set a new record for the world's highest skydive. And in the interest of making your record tough to beat, let's say you used a diving board on the International Space Station (ISS) as your launching point, 249 miles above Earth.

To get started, you would need a space suit and some

---

*Earth's atmosphere doesn't stop at any one point; it just gets thinner and thinner the higher you go. At 62 miles (100 kilometers) there's still some atmosphere, but at this altitude an aircraft has to be going at orbital speed to stay aloft—and that seems to be as good a definition as any for space.

oxygen to keep you alive for the moment (refer to p. 43 for what happens if you don't have these things). Your first challenge upon leaving the space station would be getting to where you wanted to go. You would be falling toward Earth, just like the space station, but, also like the space station, you would be traveling sideways at 5 miles per second. In fact, you would be traveling so fast sideways that as you fell toward Earth you would miss it. This is called orbit. A little confusing, but think of it this way: Consider Earth had no mountains or air resistance and we fired you out of a cannon so that you skimmed over the planet 6 feet high going 5 miles per second. Gravity would pull you down those 6 feet, but in that time you would have traveled far enough that Earth, because it's a sphere, had also fallen 6 feet.* The ISS is doing the same thing, just much higher.

Once you left the diving board you would not need any help falling to Earth. Gravity would already be taking care of that. What you would need is help decelerating so that you would stop missing the planet as you fell. So let's give you some rocket boosters to slow you down, like a Soyuz spacecraft does on its return to Earth.

As your speed decreased, you would hit that arbitrary 62-mile marker above Earth. At this point you would be falling at a blistering Mach 25. The fastest manned aircraft was the experimental X-15—basically a rocket with a cockpit. It

---

*Which means if Columbus were right and Earth were flat you could never orbit it. Also, 5 miles per second is faster than any bullet, but on the moon you would need to go only 0.7 miles per second to orbit it, which is slower than a bullet from a Swift rifle. So if you fired the Swift rifle on the moon, the bullet could circle around and hit you in the back of the head.

topped out at Mach 6.7, only it couldn't maintain that speed for long because the plane started to melt.

You would be going a few times faster. Mach 25 isn't quite man's all-time speed record—that's the Mach 32 that the *Apollo 10* module hit as it returned to Earth—but it's darn close, and Thomas Stafford, John Young, and Eugene Cernan were inside a vehicle with a heat shield when they did that. You wouldn't be.

That poses a few problems. Mach 25 is just over 19,000 miles per hour. While it's fine up at space station altitude, where there's hardly any atmosphere, as the air thickens up you begin to slow down.

That slowing down process would be painful, because the air simply couldn't get out of your way fast enough. This brings up a number of issues, but we'll focus on the big three.

The first issue is the g-forces problem. You would be slowing down so quickly, you would temporarily "weigh" 4,500 pounds. U.S. Air Force officer John Stapp proved you can withstand 46 g's for a brief moment, but 30 g's applied over many seconds—as you would experience—would be certain death. Your softer parts, like your airways and lungs, would be crushed under the g load.

The second and simultaneous issue you would experience is the turbulence problem. At Mach 25 the wind is moving so fast it would spin you around and rip you apart. When a satellite is allowed to slow down and fall out of orbit it doesn't fall in one piece; it falls in *many* pieces. And that's a satellite, which is welded metal—its limbs are attached far more strongly than yours. Even rocks are ripped apart as they make their way to Earth.

The third issue is the heat problem. All that air that couldn't get out of your way fast enough would get compressed, and compressed air gets hot. The SR-71's wings get to 600 degrees, and that's only at Mach 3.

At Mach 25 the air is hot enough to melt rock. To withstand this heat, space shuttles use tiles made of rock strands with a high melting point and such poor thermal conductivity they can be heated in a 2,200-degree oven and touched with bare hands.* Damage to the space shuttle *Columbia*'s heat shield allowed hot compressed air to enter the inside of the spacecraft, causing it to disintegrate during reentry.†

You would not have the benefit of a heat shield, so you would bear the brunt of it. The heat would carbonize your flesh, at first cooking, then burning when there's enough oxygen, and finally vaporizing you at more than 3,000 degrees.

Vaporizing is another way of saying your molecules are broken apart into separate atoms so that you become a CHON (carbon, hydrogen, oxygen, and nitrogen) gas. But eventually even the atoms of this gas wouldn't withstand the temperature.

The heat would tear the electrons from your atoms, turning you into a falling, glowing plasma.‡

The good news is your last moments would be spectacular. From Earth you would appear as streaking flames across

---

*There is a great demonstration of this on YouTube.

†Falling foam that protects the shuttle's supercooled fuel tank fell off on lift-off and put a hole in the heat shield. When the new space transportation system is introduced, it will no longer have the fuel tank higher than the shuttle in order to mitigate this issue.

‡Sorry, we don't think any part of you would remain. Ice blocks falling to Earth weighing more than two tons are completely burned up in the atmosphere, and you don't have much more durability than ice.

the sky, visible during the day and far brighter than any shooting star.

Like a typical shooting star, no piece of you would make it to Earth, at least at first. Instead, you would waft about the atmosphere as separate bits of ionized plasma.

Eventually, though, your lonely nuclei would pick up replacement electrons, become whole again, and sprinkle down to complete the highest skydive in history.

Then, because you have so many atoms in your body, after they have time to coat the atmosphere at least one of them will be in every breath everyone ever takes. Forever.

What Would Happen If . . .
You Time Traveled?

Throughout most of its history Earth has been a very inhospitable place. Either too hot, too cold, or just right but filled with terrifying predators. But let's imagine you had a time machine and wanted to see for yourself. Here's what we think would happen to you if you traveled back to . . .

4.6 billion years ago: Earth is just starting to form, but it isn't there yet. You would step into a cloud of gas and dust collapsing together under its own gravity. There's a lot of junk flinging about, some going slow enough to bounce off you and some rocks speeding by many times faster than a bullet. If you were hit by one of those it would pass clean through you. That's unlikely, though. The real problem is that Earth is still a big, unorganized pile of space trash without a surface or an atmosphere, so you're in a vacuum—expect to pass out in fifteen seconds and die of asphyxiation within a few minutes.

Earth is under construction—check back later.

4.5 billion years ago: Earth has a surface now! Unfortunately for you, that surface is made of lava, so before you have the chance to asphyxiate you would be burned alive. There aren't any solid rocks yet; everything is still molten, and nothing has cooled off. Earth also has an atmosphere, but that atmosphere doesn't have any oxygen. Not that you would have time to care about that because, again, you're standing in lava. The air does have a lot of helium in it, though, so your final screams would come out as high-pitched squeaks.

You have popped up right in the middle of the aptly named Hadean era. Better luck next time.

4.4 billion years ago: This is a slightly better time to visit, as the surface of the planet has cooled by now. The oldest rocks ever discovered come from this time period—so we know that at least there would be something to stand on.

Unfortunately, Earth still doesn't have an ozone layer to block the sun's ultraviolet light, which means you would get enough ultraviolet radiation to give you a sunburn in fifteen seconds.

And then there's the oxygen problem. Namely, there isn't any. So you would suffocate. We recommend holding your breath because: one, you might buy yourself another minute; and two, because the air is full of methane, sulfur dioxide, and ammonia—if you try to breathe your last memory will be the stench of rotten eggs.

3.8 billion years ago: Now you can actually take a swim before you die!

During the early years the solar system was a messy place with chunks of rock careening about. Earth was under constant bombardment. These meteoroids brought presents, though, in the form of new gases, and those combined with gases from the Earth's crust created an atmosphere, then rain and oceans. By this point, Earth has even wicked itself clean of the smelly sulfur—so everything wouldn't stink.

Life has also begun, so at least you will not die alone. Cyanobacteria microbes now inhabit the Earth.

There's still no oxygen, though, so you would suffocate, or, if you were really unlucky, a meteor would crush you, fry you as it passed overhead, or drown you in the resulting tsunami.

1.4 billion years ago: Something to breathe! Small organisms have been living in the oceans for more than a billion years, but recently a new guy showed up with a neat trick. This unnamed blue-green algae fed off the atmosphere's abundant carbon dioxide, then released oxygen as a waste by-product. Armed with the new technique, called photosynthesis, the algae enjoyed huge success, and over the course of a few million years changed the composition of the entire atmosphere.

Unfortunately, every other organism was doing just fine in the old atmosphere. For them oxygen was toxic, so nearly all of them went extinct in Earth's first great pollution event.

But what was bad for them is good for you. Unfortunately, the atmosphere is only 4 percent oxygen, and unless you are a Himalayan Sherpa you are used to 21 percent. Breathing 4 percent oxygen is like breathing at 30,000 feet, which is possible but requires training. So spend some time in the Himalayas before your trip.*

If you can manage the oxygen issue there is fresh water in rivers to drink, but no animals to eat and no plants bigger than algae. On top of which, if the algae is like its modern progeny (hard to know for sure), it has cyanotoxins, which are some of nature's most powerful neurotoxins. If you ate them they would paralyze your intestines and diaphragm and you would suffocate.

In other words, visiting Earth 1.4 billion years ago means death by starvation if you avoid the local cuisine, and suffocation if you don't.

500 million years ago: Your chances of survival depend on where you pop up, and beachfront property is the way to go. Nothing has crawled out of the oceans yet, so land is totally barren, but the oceans are thriving. If you appeared somewhere along the coast you would have a chance.

There's enough oxygen in the air now that you could breathe for more than a few minutes, and there are shelled organisms to eat. But be careful in the water: There are bigger fish out

---

*Other issues with life in this low-oxygen environment: If you cut yourself on your trip it wouldn't heal because your body needs energy to heal itself, and without sufficient oxygen you can't produce enough energy to fix your wounds, and giving birth is impossible because there isn't enough oxygen for a pregnant woman to share with her baby.

there, and without a shell you would look like a nice pork roast. Supersize leeches are also around, able to drill into your side and suck out your innards.

The ozone layer is still in development, so you would need to bring some industrial-strength sunscreen (something on the order of SPF 250) and a good pair of sunglasses (the UV will burn your corneas in fifteen minutes if you go without), but, all in all, you would finally have a chance at survival.

450 million years ago: The ozone layer is complete, so you would be able to venture out without getting a lethal sunburn. Sea life is booming and the rivers have fish, so you could survive. Still, there's nothing taller than a shrub, so shade is hard to come by, and finding food on land would be difficult.

370 million years ago: This is the late Devonian period—and for a time traveler interested in staying alive this might be the sweet spot. There's life on land, and there are trees to sit under, potentially edible plants, and no animals big enough to eat you. Insects are still 70 million years from showing up too, so that's nice.

You would have to get the timing just right, though, because it's starting to get a bit chilly. While trees are plentiful, there aren't yet any organisms that cause dead trees to rot, so they never return their CO_2 to the atmosphere. A good CO_2 balance is important for Earth's temperature—less of it reduces our atmosphere's greenhouse effect and produces the opposite of global warming: an ice age.* Lucky for you, the next one is still a few hundred thousand years off.

This era is our pick: air to breathe, food to eat, trees to sit under, and no mosquitoes.

300 million years ago: A huge pulse of oxygen in the atmosphere (up to 35 percent, compared to today's 21 percent) results in enormous insects.† We're talking predatory dragonflies the size of seagulls, eight-foot-long centipedes, three-foot-long scorpions, and huge cockroaches.

Bad time to visit if you're not into bugs.

250 million years ago: You have really poor timing. Fifty million years before or after this would be perfect, but at this moment 96 percent of all sea life and 70 percent of all animals are dying in the greatest extinction event on record. Earth will take 10 million years to recover its biodiversity.

Scientists aren't exactly sure what caused the die-off. One possible explanation is that many enormous volcanic eruptions—called a flood basalt—covered an area the size of India in lava and released enough CO_2 to change the composition of the atmosphere.

Whatever the reason for the die-off, mass extinctions

---

*The trees that didn't rot now make up the coal we burn for electricity, so the CO_2 that wasn't released earlier—and caused an ice age—is now being put into the atmosphere and causing global warming.

†Insects take in oxygen through their skin (aka their cuticle), so the ratio between their surface area and volume cannot be too small. With more oxygen that ratio can get smaller, and the result is dog-size scorpions.

always hit the top of the food chain hardest, and that's where you sit. There would be nothing to eat, and if the volcano theory is correct there might be something funny going on with the air.

Death by giant volcanoes. Sorry.

215 million years ago: The first dinosaurs have arrived and will wander the globe for the next 150 million years. This is a dangerous time to be a relatively unathletic human.

The Tyrannosaurus rex won't evolve for another 148 million years, but that doesn't mean you're out of the woods. Giant crocodiles, called Postosuchus, roam about and would be more than happy to eat you, as would an oversize hyena-type dinosaur called a Coelophysis.

Fortunately, nearly all the predators you need to worry about would be focused on prey living on the ground. The pterosaurs and pteranodons flying around are more interested in smaller animals, so if you spend as much time as possible in trees you would have a shot.

There are vegetation and fauna in this time period, but it's a little different from the present day. Flowers are still a few million years from showing up, for instance, so everything would look a little dreary.

For food you would be able to catch fish, spear small animals, and steal eggs for protein—just keep a sharp eye out for the parents.

There are some plants to eat, but that comes with a few caveats. Namely, some of them are poisonous. So when in doubt, follow the universal edibility test, which can be summarized as follows: Eat only one part of any plant at a time. Do not eat too much of it, and if you don't feel well, start throwing up as fast as you can.

Bottom line: If you are quick on your toes, cautious with the local food, and build a really nice tree fort, you have a chance.

65 million years ago: Avoid the Yucatán—there is a large space rock headed for that corner of Mexico. (For more details on death by meteorite, see p. 27.) In fact, you should avoid this period altogether, because the meteorite fallout will eventually kill you even if you are on the far side of Earth.

3.2 million years ago: This is the time of Lucy, the world's most famous predecessor to Homo sapiens. We know that by now our ancestors have begun to come out of the trees. That's both good and bad for you. It's good because we know man could survive in this environment; it's bad because early man could very well be the one to kill you. Lucy was shorter than you but considerably stronger—you would be a heavy underdog in a one-on-one fight.

Not to mention that you would still be in the middle of the food chain, thanks to large predators like the saber-toothed tiger roaming about. Lucy and her ilk were able to survive by grouping up, but that probably won't be an option for you.

So be nice to your fellow humanoids. Their help is probably your only hope.

If your time machine goes forward as well as back, perhaps you would like to take your chances on distant future scenarios.

Despite what you always hear, we actually do have a good idea of what the future will hold. It's not good. Here's what would happen if you traveled forward to . . .

1 billion years from now: The sun is very slowly getting hotter. Why? As the sun burns all the hydrogen fuel at its core, the nuclear reactions start moving to the surface, where there's less pressure on the explosions, and the sun expands. Even though the surface is slightly cooler, there's a lot more surface area and so more heat blasting Earth.

This is hard to notice on a day-to-day basis, but over the course of 100 million years it does make a difference. In 1 billion years, the average temperature on Earth will be 115 degrees (right now it's 61 degrees), and so hot the oceans will have boiled off.

If it were dry heat you would last hours at 115 degrees, but because all of Earth's water has boiled off, it will be extremely humid.

In other words, Earth would be a giant humidifier and you would last only a few minutes on it.

5 billion plus years from now: The sun has grown so large it has gobbled up Mercury, so the sunsets are extraordinary. Unfortunately, you would have only a few seconds to experience one.

Right now if you hold your hand at arm's length you can cover the sun with the tip of your pinkie. In 3 billion years you would need to hold up a watermelon at arm's length to block the sun. In 5 billion years it would fill the sky, which wouldn't bode well for you.

7.5 billion plus years from now: Perhaps the prettiest images in the universe are planetary nebula, which happen when a dying star throws out shells of gas that burn in gorgeous, fiery displays.

But just like fireworks, planetary nebula are best enjoyed from a distance, and if you were on Earth when the sun had its final call, you would be far too close.

Beautiful, but deadly.

What Would Happen If . . .
You Were Caught
in a Human Stampede?

Isaac Asimov calculated that if the human population continues its exponential growth, in a few thousand years we will be part of a packed ball of human flesh expanding into space at the speed of light. Exciting, but there is a problem with that theory, and it is the same problem that might come up at your next rock concert: human stampede.

If you hear the words "human stampede" you probably imagine herds of people running around like a bunch of wildebeests charging across the African savanna, but it turns out this isn't at all how human stampedes look or what makes them dangerous. In fact, the really dangerous stampedes are not when people are running—it's when they can't move at all.

Stampedes—more aptly named crushes—are typically crazes and not panics, which means a crowd of people is moving toward something they want, and not away from something they don't. If you are stuck in one you will face a couple of problems. Your first? A lack of pheromones.

In dense crowds things start to get dangerous because we, as a species, have an issue with crowd movement. Unlike ants, we're not designed for it. When ants go marching, an ant at the front of the pack can release pheromones to communicate with the ants in the back. If the way is blocked, these pheromones tell those in the back to go another way.

You don't have these pheromones. If someone trips, you can't tell the back of the pack to stop like the ants can.

In large, dense groups this lack of crowd communication becomes a serious problem. What makes a crowd large and dense? When it comes to size, if the crowd is even large enough to be called a crowd, it's large enough to kill you—but we'll get to that later. The more important factor is density. Crowd density is measured by people per ten square feet.

Ten square feet is about the same area of the chalk outline the police draw around a murder victim. The number of people stuffed inside each imaginary chalk outline, averaged throughout the crowd, is its density.

If there's an average of two people, that's a solid crowd, but walking is easy and there's little bumping. If it's double that it's called a thick crowd—there's a lot of bumping and shuffling, but people can still move.

Six people per every ten square feet is getting dangerous—you're always touching the people next to you and moving is nearly impossible.

Seven people per ten square feet is like stuffing twenty-one people into an average-size elevator—this is rush-hour-on-the-Tokyo-subway dense. In deadly crushes crowd density is typically in this zone.

At this density a crowd's movements stop resembling people and start looking like a fluid. Powerful waves that originate with people pushing in the back and gain momentum as more people are swept up pass through crowds—waves capable of sweeping you off your feet and depositing you wherever they happen to be going. If the person next to you falls, there won't be anything to hold you up and you would fall too, leading to a domino pileup of people on top of you.

If you happen to find yourself in one of these crowds—typically a religious festival, sporting event, or concert—the transition from friendly bumping to crushing can happen quickly. Suddenly you would realize you cannot raise your arms, cannot escape, and are at the mercy of the crowd.

Falling is of course dangerous, but you don't necessarily need to fall to be in trouble. Even if you're on your feet, opposing waves can pass through the crowd and pin you in place, squeezing you between the two forces in the crowd. Because of how force scales in crowds this gets dangerous quickly.

The average person can typically push with a maximum of fifty pounds of force. If there are only four or five people pushing on you, like in an overcrowded elevator, that's uncomfortable but not dangerous. In crushes people are usually not pushing with maximum effort, usually with only five or ten pounds of force each, but in a crowd of thousands, this force scales and can put a lethal strain on your diaphragm.

You need to expand your chest a few inches to breathe. Fortunately, your diaphragm is strong. A healthy person can breathe with four hundred pounds on their chest for two days before tiring.* Unfortunately, in crushes the diaphragm can be overpowered. In the aftermath of crushes investigators have found steel barriers designed to withstand thousands of pounds bent in half.

We said crowds of seven people per ten square feet can become lethal, but that is an average number for the entire crowd. At the crush point, where you're likely to be killed, there will probably be at least ten people per ten square feet. Getting that many people to squeeze that tightly is not possible without extraordinary force. It's like squeezing twenty-eight people into an average-size elevator—not possible with only a couple of unwanted passengers pushing. You would need either thousands of people pressing from behind or a bulldozer.

If you're caught between two opposing waves in the crowd, or if you fall and six or more people domino on top you, it would be like being a passenger in that overcrowded elevator with a bulldozer pushing from the back. A thousand or more pounds would press on your diaphragm and you would not be able to breathe even once.

You can replicate a thousand pounds on your chest by going three feet under water and trying to breathe through a straw. But we'll save you the trouble: It's not possible. In the

---

*We know this because in 1692 in Colonial America, Giles Corey was accused of witchcraft and pressed to death with four hundred pounds of stones on his chest. It took him two days to suffocate. His last words? "More weight."

crush, or under water, with a thousand pounds on your chest you would pass out in fifteen seconds. If it's applied for longer than four minutes you would suffer permanent brain damage and then death.

So Isaac Asimov was wrong. We know from crushes that no one can survive with six or more people lying on top of them, so Earth's population will never have the chance to be a ball of people thousands deep expanding into space at the speed of light.

The stack of people would never get deeper than six.

What Would Happen If . . .
You Jumped into a Black Hole?

Astrophysicist Neil deGrasse Tyson believes that jumping into a black hole is the most spectacular way to die in outer space. Considering that there are quite a few ways to die in space (actually, there are *only* ways to die in space, the truly spectacular thing would be to find someplace where you would not), this is saying something.

So what exactly is a black hole? In a nutshell, here's how one is created:

1. A black hole begins as a star ten times bigger than our sun.
2. Eventually the star burns all its fuel. This takes a while.
3. Without any nuclear reactions happening in the star's center, the star can no longer resist its own gravity and the outer shell collapses at a quarter the speed of light.

4. If you happen to observe this collapse, you need to run. It takes a couple of hours for the shock wave from the shell hitting the iron core to reverberate back to the surface. Once that happens the star explodes, and in that instant emits as much energy as an entire galaxy of 100 billion stars.

5. After the explosion what remains of the star collapses under its own gravity and you're left with something very small (about the size of San Francisco) but with an enormous mass (five times more massive than the sun). Its gravitational pull is so strong, its escape velocity is faster than the speed of light. This is a black hole.

So what would happen if you jumped into it?

First of all, you should know that your decision to jump is final. To exit a black hole you would have to cross the event horizon, and to do that you would have to go faster than the speed of light, which is impossible.

So far the fastest thing we have ever built is the unmanned *Helios* spaceship. It reached 157,078 miles per hour when it did a slingshot around the sun, which is fast, but only 0.0002 the speed of light. Unless you can figure out a way to go faster than Einstein says is possible, your death in a black hole is inevitable.

How you die, however, will depend upon the type of black hole you jump into. Your first option is diving into a small stellar mass black hole.

Here's what would happen: Once your feet left the spaceship you would begin free-falling into the hole. This would not be your typical free-fall, however. By the time you reached the event horizon of this stellar mass black hole you would be falling just short of the speed of light, or roughly 186,000 miles per second.

Interestingly, you would be fine. Normally, traveling through space at the speed of light is not recommended. It's not the speed or acceleration that's dangerous, though; the issue is hitting stuff. Even teeny particles pose a big problem when you're going that fast—and space is not a perfect vacuum. It is littered with bits of hydrogen that hit like atom-destroying bullets when you're traveling near the speed of light. The hydrogen would smash through your body and destroy the nuclei of your atoms, which would be fatal.

Most black holes are surrounded by a pure vacuum, so you shouldn't hit enough hydrogen to kill you—just make sure you don't jump into a messy black hole that's surrounded by orbiting gases.

If you chose correctly, your acceleration to near light speed should pass smoothly. As you got closer to the black hole, though, you would feel your body beginning to stretch as the strength of the gravity increased so dramatically the pull at your head (assuming you have maintained pike position) would be stronger than the tug at your feet, stretching your head away from your toes.

At first the pulling would feel good, like a gentle tugging at the chiropractor. However, it would quickly get uncomfortable and you might begin to realize the trouble you were in.

This "tidal force" on your body would eventually rip you apart as if you were tied to two trains moving in opposite directions.* First you would be pulled apart where you're weakest, right near the belly button, where you have only a spinal cord and soft fleshy parts. There aren't any vital organs in your lower half and bleeding to death takes a while, so you would still be alive. For the moment. But the tidal forces increase as you continue toward the center, so you would rip again. And then again and again. Your body would continue to split until you were nothing but a head speeding toward the singularity. And then that would be ripped apart too.

This dicing would all happen very, very quickly. Someone would need to film you in slow motion to see what actually happened. To the naked eye you would just be gone.

But it gets worse. The black hole's gravity not only pulls your body but also constricts it, squeezing you like the ultimate corset. Eventually the gravitational forces would become stronger than the chemical bonds in your body, so the gravity wouldn't just split your body into pieces but your molecules as well—stringing them out until you were nothing but a parade of atoms speeding toward the singularity.

Black holes do not allow light to escape, so there's no way to know what the hypothetical singularity looks like, or what *you* would look like inside. However, wherever you are in the black hole and in whatever form, we know it is not your final resting place. A black hole gradually leaks out Hawking radiation until it evaporates entirely. So at some point many tens

---

*In medieval times this was known as drawing and quartering, and they used horses in place of trains. But horses are not as strong as black holes, and sometimes were incapable of pulling a man apart—needing assistance from the headsman's ax.

of billions of years later your remains would reappear outside the event horizon in the form of a few radiated photons.*

But let's go back and allow you to change your mind. Let's say instead of jumping into a small black hole, you leapt into a supermassive one. Your death would still be a certainty, sadly, but it would also be a little more interesting.

Since the gravity of these supergiant holes increases more slowly, you would actually make it beyond the event horizon alive. What happens after that is a profound mystery. Because light can't escape from inside a black hole, it is impossible to know what happens inside. We cannot look in because light doesn't bounce back, and any probes that crossed the event horizon would vanish. No signal would ever emerge.

But we can still speculate. The manner of your death would likely be the same—stretched out and spaghettified by the tidal forces from a singularity at the center. But because you would still be alive inside the horizon, your final moments would be a little different. You could look out at your friends on their ship, only your vision would not be perfect. Everything would appear distorted, because not only would you be bent and squeezed into a black hole but light would be

---

*Here's where things get really complicated. We said earlier that you can't escape a black hole unless you go faster than the speed of light. And that's true. Thanks to Einstein, we also know that nothing with mass can go faster than the speed of light. So now we have a contradiction. How is it that you're able to escape the black hole as radiation? The answer: the same way you get a file off a flash drive. Electrons in the flash drive are stored in energy wells; they enter and leave the wells by a quantum mechanical process called tunneling, in which they vanish from inside the well and appear outside it without passing through the space in between. In the same way, particles can vanish from inside a black hole and reappear outside without crossing the event horizon. So the bad news is if you jump through a black hole, your atoms will be shredded apart. The good news? You will learn to teleport.

too. So looking out at the galaxy would be like peering at the stars out of a porthole, or looking up at the world from under water—the stars and planets would all be compacted into your tunnel vision.

And then you're (spaghettified and compacted into sub-atomic particles) dead.

What Would Happen If . . .
You Were on the *Titanic* and Didn't Make It into a Lifeboat?

Lᴇᴛ's sᴀʏ ʏᴏᴜ were one of the lucky 2,228 people to earn passage on the maiden voyage of the RMS *Titanic* in 1912. You gladly paid the three hundred dollars (what it would be worth today) for your third-class ticket and the privilege to ride two floors below Europe's elite to America.

As you may have heard, this trip ended poorly.

Once the boat hit the iceberg, passengers had just over two hours to find lifeboats, of which there weren't enough. In third class, fewer than half of the women survived and just 16 percent of men made it out alive.* In third class, you probably wouldn't have made it onto a lifeboat, and instead would have fallen into the North Atlantic Ocean. What would happen next?

---

*A first-class ticket on this voyage, though pricey (almost two thousand dollars today), was worth it on this trip: 97 percent of the women and 32 percent of the men in first class lived.

The ocean's salt allows the water temperature to drop below its freezing point. In the North Atlantic, where the *Titanic* went down, the water was 28 degrees, and because water is already extremely efficient at lowering your body temperature by virtue of being so dense, you would be in one of the world's most dangerous places to go for a swim. You would be rubbing up against molecules packed 800 times denser than they were just a few minutes ago on the *Titanic*'s deck, which means you would cool down 25 times faster in 28-degree water than in 28-degree air.

Your first reaction to that rapid cooling when you hit the water would be a gasp. If your head was under water you would risk taking water into your lungs, which is dangerous regardless of its temperature, so you would need to keep your head above the surface for the first part of this (and really for all the parts if you could help it).

Your second sensation, other than a chill, would probably be a headache. One of life's early lessons comes in the form of a headache. When you drank your first milk shake, you probably did it too quickly and froze your brain. Or at least that's what it felt like. What you really did was freeze a nerve that runs along the roof of your mouth. When that happened, your brain reacted—or actually overreacted. It thought your whole head was freezing, so it diverted extra-warm blood to itself, which caused it to swell and created a size problem: too much brain and not enough skull. The result was an ice-cream headache.

That same thing would happen when you first hit the water (although in this case your brain wouldn't be fooled into thinking it was freezing; it would *actually* be freezing). Your brain would receive a rush of warm blood, swell up, and give you a blistering headache. Then you would spend the next thirty seconds in cold-water shock and start hyperventilating.

Prolonged hyperventilation gets rid of too much CO_2 from your blood, dropping its acidity. If your blood's acidity dropped too low you would faint, and losing consciousness while swimming is a bad thing.

If you could manage to stay conscious, the next thing you would undergo is an attack of muscle spasms—called shivering. Shivering is your body's attempt to warm itself up by engaging its muscles. Basically, if you're not going to do jumping jacks, your body will do them for you. Unfortunately, shivering makes muscles bad at what they do—coordinated motion. That's fine if you're waiting for the heat to come on in your house, but in freezing-cold water you need your muscles to get you out of the pickle you're in, which you can't easily do when they're twitching and shaking uncontrollably.

Both the shock and the shivering are parts of an overreaction, a misfiring of your body's fight-or-flight response, which has evolved to keep you alive. Repressing them is possible with training, but even if you have trained yourself to suppress your body's overreactions, there are a few physiological changes you can't avoid.

For one, your arteries would shrink so much your heart would have to go into overdrive in order to force blood through them. Meanwhile, your brain would be reassessing priorities and diverting warm blood away from your limbs and to your critical organs.

Your extremities would go numb because the chemistry of your muscles and nerve fibers works best at body temperature. As your nerves cool, your muscles would lose their strength and your limbs would lose feeling. Basically, your toes would freeze because your brain was throwing them under the bus.

The numbness in your hands and feet would creep higher, so that after 15 minutes at subzero temperatures, your arms and legs would lose feeling. This is bad for swimming. Most people who die in cold water don't technically die of hypothermia. They drown. Which is exactly what you would do at this point without a life jacket.

The good news is, if you have some flotation, you can survive a surprisingly long time. Even in freezing water.

That's because not only is your flesh a good insulator, you're also really good at generating heat. Right now you're using that heater to keep your core body temperature at 98.6 degrees. Once you hit the ice water that number would start to drop, but a bit slower than you might think. You would have between 30 and 60 minutes (depending on how much insulation you have) before your temperature dropped to 90 degrees. At this point you would fall unconscious. This isn't great when it comes to swimming, but assuming you have some flotation and your head is above water, you would still be alive.

Sometime 30 minutes after falling in, you would progress beyond moderate hypothermia. Stay in much longer and things get dangerous. After 45 to 90 minutes your body would reach 77 degrees and you would go into cardiac arrest. Normally that means you would likely die. But in this case, you may still have a chance. Your heart is a little like your car's dead battery—it *can* be jump-started. The part you really have to worry about is your brain—once it doesn't have any electrical signals it's gone for good, and for reasons that aren't well understood, your brain cells don't need as much oxygen when they're chilled.

Whenever people go in for risky heart surgery, as a safety measure the doctors first cool them down. If something goes wrong and the patient's brain stops getting oxygen, the cooling gives the doctors a little buffer time to fix the problem. With a low body temperature your brain can go as long as 20 minutes without air before it starts dying. Under normal conditions you have only 4 minutes.

The record holder for returning to life from the frozen dead might be Anna Bagenholm, a Swedish skier who fell through thin ice and became trapped. Anna found a pocket of air, but after 40 minutes in the water she went into cardiac arrest. By the time she was rescued—another 40 minutes after her heart had stopped—her body temperature was 57 degrees. Despite all that, after 9 hours of resuscitation she made a full recovery.

So the cold kills you at first, but in the end it could be what saves your life, and it's why doctors say you're never dead until you're warm and dead.

You Were Killed by This Book?

Sᴏᴛᴛɪɴɢ ᴛʜᴇʀᴇ ʀᴇᴀᴅɪɴɢ this book you might not think you are holding a lethal weapon. You probably think you have never seen a less lethal object in your life, but that's where you're wrong. If you were to properly employ this book's kinetic, chemical, or nuclear energy it could destroy you, the bookstore, or your entire city. How do you turn this book into an instrument of gruesome lethality? Let's start with *And Then You're Dead*'s kinetic energy.

Dropping this book won't make it lethal. Even if you were reading this on top of the Empire State Building, it wouldn't build up enough speed to do any damage.* Its terminal velocity is only 25 miles per hour—slower than you could throw it.

---

*That's not true for all books, however. The second edition of the *Oxford English Dictionary* weighs 172 pounds, and if it was dropped from the Empire State Building its terminal velocity would be 190 miles per hour. That would crack your skull and snap your neck.

And we'll stop you right there. Throwing it won't do the trick either. A 50-miles-per-hour book might hurt but is definitely not lethal.

But what if you launched it from a book cannon?

At 100 miles per hour this book would hit you with roughly the same force as a baseball, which would hurt but most likely not kill you (though a 100-miles-per-hour baseball has killed before). So let's take it up a notch.

A copy of *And Then You're Dead* (*ATYD*) hitting you at the speed of sound would penetrate your skin and knock you down. You would probably survive if it hit you in the arm or leg, but if it hit you in the chest the shock wave could disrupt your heartbeat and kill you.

If we sped the book up to Mach 10, it would hit you with 5,000 times the energy of a 100-miles-per-hour copy. The book would compress and heat the air in front of it so that it would fly toward you as a 3,000-degree incandescent ball. Unfortunately for you, it would not burn up entirely. It would if you just left it there—it's certainly hot enough—but it isn't just lying there. It's traveling toward you at 10 times the speed of sound, so it doesn't have time to burn up. Instead, it would embed itself in your chest as a 3,000-degree paper cannonball.

But let's fire it faster. Mach 200 is the fastest a man-made object has ever traveled. To get the book up to this speed you would need to build a giant potato cannon with a nuclear bomb functioning as the hair spray.* At this speed the book is a

---

*While the average potato cannon is powered by burning hair spray, the greatest potato cannon ever made is also known as the Bernalillo underground nuclear test, which occurred in Los Alamos, New Mexico, in 1957. The U.S. military set off a

flying plasma sphere coming toward you at more than 150,000 miles per hour. It would take 1 minute and 12 seconds to travel from New York to San Francisco. If it hit you, you would be blown apart in a big mess of body parts and pages.

That's using this book's kinetic energy, but to do even more damage you should take advantage of its chemical properties.

Putting a match to this book will barely warm your hands. But that's not making the best use of its potential chemical energy. The best thing to do is the same thing scientists do when testing the number of calories in a candy bar: Explode it.

The way scientists test the calorie content of food is to dehydrate it, grind it up, and place it in a pure-oxygen-filled steel container, then spark it. The power of the explosion (equal to approximately one stick of dynamite, in the case of the candy bar) is the measure of the food's calories.

A copy of *ATYD* contains 1,600 calories,* or nearly a full day's worth of food if you were like a termite and could digest paper's cellulose. If you ground this book up, put it in a steel container with pure oxygen, and sparked it, it would explode with the same power as five sticks of dynamite.† If you were

---

smallish nuclear bomb underground and had a high-speed camera trained on a large manhole cover that covered the well leading all the way down to the bomb. The camera took 160 photos per second, yet caught only one shot of the cover before it disappeared out of frame—meaning it was traveling at an absolute minimum of 41 miles per second.

*Note that 1 food calorie is equal in energy to 1,000 theromodynamic calories, but in this chapter we're referring exclusively to the food version.

†This is illegal, incidentally, as is any firework weighing more than 3 grams of gunpowder. Which means that legally the most of *ATYD* you're allowed to grind up and explode is this page.

reading it at the time, that would certainly kill you. But we're still not getting the largest possible explosion out of this book.

If you're looking for a bigger boom, you will need to release *ATYD*'s nuclear energy.

All mass has energy: This book. Your coffee mug. The chair you're sitting in. Everything. And when you convert mass to energy, you get big numbers very quickly. The atomic bomb that exploded over Nagasaki converted a single gram of mass (equivalent to less than half a page of this book) into energy. The trick is making the conversion happen. Fortunately, it's not easy to do. The Nagasaki bomb used plutonium because plutonium is unstable and easily converts to energy. Books like *ATYD* are far more stable.

So it's difficult to convert this book's mass into energy—but it is not impossible. The best way to accomplish it is to create a book of antimatter and combine it with your copy.* Then back away. Quickly.

Release this book's nuclear energy and it would explode with the power of the largest hydrogen bomb the United States has ever detonated. You would get so hot, each of your atoms would break off, then your atoms' electrons would be ripped off, and you would be scattered about the atmosphere as ionized plasma.

Creating that much antimatter is beyond our capabilities right now—the most antimatter we have ever made is 17

---

*What is antimatter? It's complicated, but suffice it to say that every atom of matter has an "evil twin" of antimatter, and when a particle of matter touches its antiparticle, both vanish and are converted into energy according to Einstein's equation $E=mc^2$.

nanograms (17 billionths of 1 gram) of antiprotons, and that took many years, so an exploding book is a problem for future generations. But there are more realistic ways you could turn this book into a lethal weapon—like turning a page too quickly.

A single paper cut could kill you. It has happened before. In 2008, an English engineer sliced a quarter-inch paper cut on his arm just before leaving on a trip to France. He soon developed flulike symptoms, became weakened with fatigue, and grew delirious. He died in the hospital six days later from necrotizing fasciitis, a rare but nasty bug that infects through even the smallest wounds and cuts.

It is a hypochondriac's worst nightmare.

Unbeknownst to you as you read this page, the necrotizing fasciitis bacteria could be living on your skin. If you're hasty in turning the page and the paper slices your finger, the erstwhile harmless bacteria could gain entry.

Part of necrotizing fasciitis's charm is that it lives within dead tissue that neither antibiotics nor white blood cells can access, and as the bacteria grows, it belches out a mix of exotoxins that kill your cells before your immune system can mount a defense. Without early intervention, you will progress beyond physical pain into severe sepsis.

Sepsis is your body killing itself in an effort to stop the invader. Your body reroutes so much blood that your heart won't be able to send any to your brain. At first you will feel faint and confused as your brain sputters along on the bare minimum. As your blood pressure continues to drop it leads to multiple organ failure, most critically your heart. Once that fails your brain stops receiving oxygen and you die within a few minutes.

Without medical attention, the death rate for necrotizing fasciitis is 100 percent. Even with early medical care 70 percent die, making it more deadly than the Ebola virus.

Be careful as you turn this page.

What Would Happen If . . .
You Died from "Old Age"?

THE MOMENT YOU were born, the odds of your death sky-rocketed, and your very first day was one of your most peril-ous. (It looks like you made it. Congratulations!) Even if you were born on time and without any congenital abnormalities, you had a 0.04 out of 1,000 chance of death, the same odds of dying in one day as a ninety-two-year-old has. As you grow older your immune system strengthens and your chances of dying decrease every day.

Your twenty-fifth birthday should be a celebration, not only because you're allowed to rent a car but also because it's the healthiest day you will ever have. You have overcome childhood diseases and your adult life begins. From here, it's all downhill.

Every day you get older, the odds of your death increase at a very predictable rate.

In 1825, working as an actuary for an insurance company

he helped found, Benjamin Gompertz published his law of mortality—every eight years after your twenty-fifth birthday your chances of death double. He discovered that humans, just like fruit flies, mice, and most other complex biological organisms, die off at an exponential rate.

What we don't know is why we die so predictably or even why we age. There are several theories but so far nothing has been proven. One possible answer is called the reliability theory.

According to the reliability theory, when you were born your body was absolutely littered with errors and failures in key components. Sorry, nothing personal, the same goes for everyone you know. It turns out that humans may be built like old French cars, riddled with faulty parts. Not only that, but the parts that do work break down all the time.

Fortunately, unlike a Renault Dauphine,* you have a lot of redundancy. Cells are small, and around 37 trillion of them have been stuffed into your body. It seems like nature knew it had unreliable parts but, unlike the Renault manufacturer, cost was no concern, so it built as many backups as it could. However, over time, as more and more of these cellular backups break down, you "age" until you're all out of spare parts, at which point you die.†

---

*According to *Road & Track* magazine you could hear this car rust if you stood close enough.

†We know a version of the reliability theory is how hearing works. Inside your ear you have little hairs to detect vibrations, along with many backup hairs. Loud music snaps them off, which isn't a problem when you're young, but as you age, the hairs die off naturally, and because rock music has already taken your backups, your hearing goes.

Of course there are ways to accelerate or decelerate this process.

Once you hit age twenty-five, you have roughly 1 million half hours left to live, give or take. Therefore, every half hour after your twenty-fifth birthday counts as one microlife. Using this as a baseline, Cambridge statisticians David Spiegelhalter and Alejandro Leiva created a way to measure the cost or benefits of different lifestyles.* For example, smoking two cigarettes costs you one microlife, i.e., your expected life-span is a half hour shorter after smoking those cigarettes. Smoke two more? That will cost you another microlife. Ten pounds overweight? That's a microlife per day. More than one alcoholic drink per day costs you another microlife per drink. Living and breathing Mexico City's air pollution costs you one half of a microlife every day.

That's the bad news. The good news is you can also add microlives to your account with good behavior. Twenty minutes of exercise? Add two microlives. Eating your fruits and vegetables? Add four microlives per day. Drinking two to three cups of coffee adds another microlife. By simply staying alive you are gaining twelve microlives per day due to medical advances.

Eventually you will run out of cellular backups and your account of microlives will drop to zero, which explains why, at least according to the latest research done by *The Onion,* the world's death rate is holding steady at 100 percent.

---

*Microlives are similar to the micromort, a concept introduced by researcher Ronald Howard—see p. 145–46 for details. A microprobability is the one-in-a-million chance of a given event occurring.

What Would Happen If...
You Were Stuck in...?

Cᴌᴀᴜsᴛʀᴏᴘʜᴏʙɪᴀ, ᴛʜᴇ ғᴇᴀʀ of suffocation or being confined, is one of the world's most common phobias. Studies suggest that at least 5 percent of the world's population suffers severely from it, but in nearly every case the fear is unwarranted. It's a gross overreaction of the body's fight-or-flight response that usually does more harm than good.

However, there are some places where you *should* be concerned if you happen to find yourself stuck, where even the most claustrophobic brain may undersell the danger. Here's what would happen if you were stuck in . . .

An Airplane Wheel Well

Since 1947, 105 people have stowed away in an airplane's wheel well. In nearly every case it was a bad idea. But if you're still on the fence about buying a seat, we'll make a pros-and-cons list of stowing away in a wheel well for you.

Pros:

1. It's cheap.
2. You can skip the Ambien. Once the plane reaches cruising altitude you will pass out from lack of oxygen and remain unconscious for the remainder of the flight.
3. Depending on the airliner you may actually have more leg room than in coach.

That's about it for the pros.

Cons:

1. The odds aren't in your favor. Of the 105 people that we know of who opted to fly wheel-well class, only a quarter survived, and most of the survivors were either young (with less body mass younger people cool down

faster, and we'll get to why that's a good thing later) or on short flights that don't fly as high (for which we would recommend a bus).

2. The cold problem. At 35,000 feet it's 65 degrees below zero outside. You will have a bit of insulation once the wheel door closes, so the cold probably won't kill you, but don't be surprised if you lose a digit or two.

3. The exposure issue. There are no seat belts in wheel-well class, and the doors open for landing when you're still at a few thousand feet. You would not be the first to fall. We would say hold on, but because of the lack of air you're going to be unconscious.

4. The lack of oxygen. This is the real killer. At 35,000 feet the air is so thin the oxygen you breathe drops to 25 percent of what you're used to. People get woozy at 50 percent, so unless you are acclimated you will pass out unexpectedly and die a few minutes later. Your best shot is to allow yourself to *nearly* freeze to death. Your brain needs far less oxygen when chilled, so wearing shorts and a T-shirt may actually be a better idea than a jacket. You will probably lose at least some of your fingers and toes to frostbite, but as long as you're not dropped you may make it out of this.

Conclusion: The trip won't cost you an arm and a leg, just some portion of your hands and feet. If you're really lucky.

A Gas Station (Or: What Would Happen If You Lived on a Diet of Only Junk Food?)

Surviving a single gas station hot dog is a feat, so how long could you live on a diet of them?

Junk food, for all its faults, preserves quite well. Potato chips last a long time, and it's unclear if a Twinkie will ever rot, so you would not starve, though if you ate junk food and soda for years you would probably develop diabetes. However, there's a more short-term problem: Junk food has practically no vitamins or minerals, so while it makes for a good snack, it's a poor meal.

If there were any fresh fruit in your gas station, it would expire within a few days. And without fruit you would get almost no vitamin C, which is one of the worst to skip if you're going without vitamins.

In the early 1500s bigger boats and better maps made long sea voyages possible. Unfortunately, food preservation lagged behind—which meant no fresh food, no vitamin C, and the inevitable result: scurvy.

Magellan lost 80 percent of his sailors to scurvy on a voyage across the Pacific, and even by 1740, thirteen hundred sailors died of it in ten months aboard a ship captained by the explorer George Anson.*

A month after you started your diet of pure junk food in the gas station, you would show the first signs of scurvy (gum bleeding, fatigue, skin spots). After another month you would be unable to repair your capillaries and you would bleed to death.

Conclusion: If you're ever stuck in a gas station, you had better hope it stocks multivitamins.

An Elevator

The record for the longest elevator ride may belong to Nicholas White, who, while working late one Friday evening in October 1999, took the elevator for a smoke break and didn't get off it for forty-one hours. The technicians apparently didn't check to see if anyone was inside before shutting it down. White spent a very boring weekend in the small box but was eventually discovered and rescued, and, according to Mr. White, in need of only a beer.

It was a good thing White wasn't stuck much longer, because elevator trappings can be lethal. In 2016 in a busy Beijing apartment complex, an elevator malfunctioned between

---

*The Royal Navy was among the first to discover the link between vitamin C and scurvy, so it gave their sailors limes to eat on their voyages—bestowing on them both a significant military advantage and a nickname: Limeys.

the tenth and eleventh floors and technicians cut its power without seeing if anyone was in it. Someone was. A month later her body was discovered.

The greatest danger when trapped in an elevator is lack of water. Elevators have good ventilation, so oxygen isn't a problem, but dehydration is. Just sitting around sweating and breathing, you lose two cups of water per day—and then there's the peeing.

Urine is 95 percent water, and after a few days stuck in an elevator, extreme thirst may make it look like a refreshing beverage, but your body is getting rid of the remaining 5 percent for a reason. It's filled with enough potassium to send you into kidney failure if you drink too much. There's also enough sodium in it to make it a poor choice for hydration.* The U.S. Army's survival handbook advises against drinking it.

As you lose more and more water through sweating, peeing, and exhalation, your blood becomes thicker and thicker, oozing through your veins until your heart can no longer pump; meanwhile, your kidneys are being poisoned from blood that's too concentrated.

Conclusion: If you get stuck in an elevator, you have about two weeks before you die of kidney failure—and pass on the pee while you're in there.

---

*If you have a soda, you *should* drink it. It has some salt so it's not as hydrating as water, but it still does more good than harm.

A Walk-in Freezer

Modern walk-in freezers are required to open from the inside, so you cannot be locked in. But what if you got stuck in an old one wearing shorts and a T-shirt?

In the 10-degrees-below-zero temperature of a meat locker, your body will reroute your blood to your core to keep your vital organs warm, which will leave your extremities out in the cold. That means frostbite, and in a meat freezer that would happen within thirty minutes. If you were able to stay alive, your fingers would eventually blacken, die, and require amputation. But you would be dead long before that became a concern.

In a meat locker your body temperature would fall a degree every thirty minutes, so after six hours your body temperature would fall to 86 degrees and your cells would stop working. Unfortunately, all you are is a bunch of cells.

Conclusion: You have only six hours before you join the meat section. According to FDA rules for similar meat (veal) you would stay fresh for four to six months before you would have to be thrown out.

Quicksand

In the enormous world of overstated Hollywood risks, dying a horrible death in quicksand might take the top spot—which, from the industry that gave us forty-foot sharks, killer computers, and alien parasites, is saying something.

Despite what you might have seen, there have been no confirmed deaths in quicksand ever. Zero. A few people *might* have been stuck in mud near the shore break and drowned in the incoming tide, but that's it.

The reason for quicksand's lack of lethality is that you float in it. It's twice as dense as water, and you already float in water. If you stepped in quicksand you would sink only up to your belly button before you became neutrally buoyant. The only way you would ever be in trouble is if you went in head-first. Avoid that and you should be fine.

Conclusion: Yes, you could conceivably die in a quicksand pit, but in all of human history you might be the first to do it.

What Would Happen If . . .
You Were Raised by Buzzards?

Hákarl IS A gourmet treat and national dish in Iceland. It is also, according to chef and experienced eater Anthony Bourdain, "the single worst, most disgusting-tasting thing I've ever put in my mouth." That's probably because the recipe involves allowing shark meat to rot for six months. That's not done to improve its taste, but rather because the Greenland shark's flesh is poisonous. If you eat it fresh, the effects from the toxins resemble extreme drunkenness. Rotting is the only cure and the final product reeks of ammonia. Apparently, it's an acquired taste.

Hákarl is one of the few foods safer to eat rotten than fresh. Most go the other way. When an animal dies on the prairie, it loses its ability to fight off infection. That's not critical to the animal, obviously—its fight is over—but it is a big deal to anything trying to eat that animal's dead flesh.

These infections produce some nasty toxins as by-products. The more recently an animal died, the fewer toxins it's likely to have.

The ultimate expired-food eater is the turkey vulture, so, disregarding the difficulty of establishing an emotional connection, let's look at what would happen if you were abandoned as a babe on the open prairie and adopted by a band of buzzards.

Food would definitely be an issue. You might have heard that playing in dirt strengthens your immune system. The turkey vulture is the ultimate example. They have been eating rotting carcasses and not washing their claws for millions of years, so they've built up a killer immune system. Because of that, their idea of a Thanksgiving feast looks quite a bit different from yours.

The first thing you might notice when joining your new adoptive family at the table: maggots.

Maggots grow from fly eggs and once they hatch they will begin competing with you to eat the rotting carcass. The good news is that maggots are a source of protein, and because they're alive they're actually safer to eat than the rotten food, so help yourself. Maggots also like to eat rotten flesh, which means the meat they leave behind will be a little fresher. So if you see them preferring one rotted carcass over another, hold your nose and dine with the maggots.*

Which brings up the second issue: smell.

---

*Maggots get a bad rap. They are sometimes used in medicine to clean out wounds, because they eat only rotting flesh and leave behind anything still alive.

Rotten food smells bad to humans for a reason. We have been naturally selected to find the smell revolting. We can detect the two chemicals that organisms produce when they die, putrescine and cadavarine, in minute amounts. Which is good—it's an adaptation that kept our ancestors alive. But it's amazing what you can get used to.

If you were raised by vultures, you would probably grow to love the scent of rotted meat. The smell of skunk is addictive to workers who handle it, and durian is a Southeast Asian fruit that smells like raw sewage—yet those who eat it regularly love it.

Smell plays a large role in taste, so we think that while the adjustment would be rough, in time you would grow to love your new rotten diet. Unfortunately, time is not something you're going to have a lot of, because your stomach and immune system would not adjust as quickly as your nose.

Eating old dead animals exposes you to whatever pathogen is eating their flesh. You could wait to see if the meat kills one of your fellow buzzards, but that's not reliable. Buzzards have a number of adaptations that allow them to eat meat that would kill you. Their stomach acid, for one, is a hundred times stronger than yours. With a pH between 0 and 1, it's stronger than battery acid and can dissolve metal. On top of that, they have the strongest immune system of any vertebrate animal. They're resistant to cholera, salmonella, and even anthrax—all lethal to humans. If you and your buzzard family dined on an animal infected with any of these, your family would be fine—but you would be dead.

However, if you were raised by buzzards, at least you would learn one good habit: A buzzard's urine is so acidic it sterilizes everything.

So after eating a hearty meal of rotten meat you might as well do as the buzzards do to clean up: Pee on yourself.

What Would Happen If...
You Were Sacrificed into a Volcano?

THE VIRGIN SACRIFICED into a volcano is, in reality, almost entirely Hollywood fiction. The cultures accused of doing this didn't have good volcanoes for sacrificing, and even if they did, hiking all the way up a volcano just to throw someone in is pretty impractical.

Still, let's say they made an exception in your case. Let's say you were pitched into a volcano. Your first question: Will you sink or float?

This may seem like a technicality, but it has some relevance to you. Not whether you would live, of course; unfortunately there's no chance of that, but it would change your exact mode of death.

Lava is melted rock, so it's two to three times denser than water, depending on its composition. It's dense enough that if you stumbled upon a river of lava you could trudge across it if you ignored the heat issue—so, yes, you would float. At least initially.

But this actually presents a problem. Sinking is a good thing when jumping from tall places into a liquid.

If you were tossed from the rim of a decent-size volcano you would sink into the lava only a few inches. The heat would be the least of your concerns. It would be like jumping from a five-story building and expecting to survive because you aimed for a sand pit. The result? Not surviving.

So hopefully your volcano has a short drop. That would give you a few more moments. Of course, that would leave the matter of the heat.

Lava is between 1,300 and 2,200 degrees. It's so hot you wouldn't even cook or burn—you would flash boil, which means all your water would turn to steam. Since you're mostly water, this is bad. Once your water converted to gas you would turn into a bubbly mess, and all that bubbling would churn and broil the lava into big lava fountains. These fountains can shoot up surprisingly high, five or six feet, and they would cover you in the stuff.

So eventually you would drop below the surface, but to be technical, it's not because you were sinking.

It's because you were being buried.

What Would Happen If . . .
You Just Stayed in Bed?

IF YOU'RE MIDDLE-AGED, you face a one-in-a-million chance of death just by getting out of bed and adding up the small risks you take every day by driving to work, cleaning out gutters, and walking over street grates. It's enough danger to make you want to stay under the covers.

But if that's your plan, think again. It turns out that staying in bed actually makes your chances of death skyrocket.

Inactivity in itself is bad for your health. In the United States it kills more people than smoking. Sitting on your butt watching a movie knocks a half hour off your life expectancy per feature, according to research by Cambridge professor David Spiegelhalter. If you did that all day every day, your life would speed by 25 percent faster than everyone else's.

But you would be dead long before that if, petrified by the risks of daily life, you stayed in bed and never got up.

Strict bed rest is extremely dangerous. It's akin to the effects of zero gravity, and part of the reason NASA has had astronauts stay in the space station for a year is to study what might happen on a trip to Mars (a seven-month one-way trip).

If you stayed in bed for seven months, like astronauts going to Mars, you would face a few issues.

After just twenty-four hours without exercise, your muscles would begin to atrophy, starting with your calves and quads, which are the ones most accustomed to a daily workout. But not only do your muscles waste away without a workout, so do your bones.*

When you switch to a horizontal lifestyle, strange things start happening with your fluids. The liquid that sloshes around your cells is used to gravity pulling it down, and if you were to lie flat for too long, these extracellular liquids would start to creep upward into your face, crush your optic nerve, and mess with your balance and sense of smell.†

Your blood is also used to gravity and exercise. During their Mars study, NASA had patients wear compressive sleeves on their legs. Why? Your veins need help returning blood from your legs to your heart. Normal walking and flexing of your muscles are usually enough, but while lying immobile, blood can pool, coagulate, and clot within a vein.

This is bad.

The pressure of downstream blood will break the clot off. It should move fine through your larger arteries, but unfortu-

---

*Your bones are piezoelectric, which means they generate electricity when stressed (just like the crystal rock in your barbecue lighter). No bone stress means no electrical signals, which means no rebuilding and results in bone disintegration.
†Wandering extracellular fluids are why astronauts returning from the space station have puffy faces.

nately your heart and brain have narrower valves and veins. The clot can get caught up in these choke points and form a dam.

If the blockage is in your heart it causes a heart attack, and if it's in your brain it can cause a stroke, either of which could kill you within a few minutes.

But heart attacks and strokes only *might* kill you, and precautions like wearing compressive sleeves can be taken. However, if you stayed in bed for seven months and weren't careful, bedsores *would* kill you.

Bedsores happen when the pressure between your bed and your bones kink your blood vessels shut, starving your skin of oxygen.

The pain would start off as a dull ache but within a few hours could progress into a painful sore. The more time you spend in bed, the worse they would get. Eventually the ulcers would progress from reddish sores to deep wounds surrounded by dead tissue.

This is when infection would become a real concern. Your skin is your primary defense against outside germs, and a continuously open wound provides a direct path for outside bacteria to enter your bloodstream and spread to your organs—sepsis.

Without immediate care—and sometimes even with it—sepsis will kill you. Your body's response to an infection is overwhelming. Your blood pressure would drop to dangerous levels, your kidneys would fail, your breathing would quicken, and eventually you would lose the ability to swallow, leading to the gurgle and crackle sometimes called a death rattle.

Eventually enough brain cells would die that you would lose consciousness and fall into a coma.

All this from just lying in bed. So if you're trying to lower the daily one-in-a-million shot at death-by-accident, allow us to propose a few better ways.

First of all, get out of bed and move away from tornadoes. Kansas, Oklahoma, and Kentucky rank as the most dangerous states for natural disasters. You will also want to avoid the northern Midwest, like Minnesota and North Dakota (too much ice), and the South (too many hurricanes). The best state for avoiding natural disasters? Hawaii. But Hawaii has a lot of two-lane roads, so their driving safety ranking is middling. The safest roads, and the safest state to live in, period, is Massachusetts. Not too many natural disasters, safer cities, and the safest roads.

You should also avoid cars. Every 230 miles you drive is a one-in-a-million chance at death. Use trains instead—you can go 3,000 miles before you get to the same risk level.

You will want to be married. That adds ten years to life expectancy.

Nursing homes are the most dangerous places to work, just edging out firefighting. The safest job? Money manager.

So if the one-in-a-million chance of death each day seems high, don't stay in bed; instead, get married, move to Boston, become an accountant, and take the train to work.

What Would Happen If . . .
You Dug a Hole to China and Jumped In?

At some point in your life, probably early on, you might have attempted to dig a hole to China. You might have even made it part of the way, maybe three or four feet, depending on the type of sand at your beach.

But now you're older and more persistent. So let's say on your next trip to the ocean you succeed where you previously failed and dig a hole all the way through the planet, all 8,000 miles. And then you jump in.

What would happen?

First, it depends on where you began digging. Exactly where you start is important. The belief that China is on the other side of America is actually incorrect. In truth, if you started your hole anywhere in the continental United States you would end up drowning in the Indian Ocean. To start your hole in the United States and emerge on dry land you would need to begin on a beach in Hawaii, where, after digging, you would surface in a Botswana game preserve.

But starting in Hawaii has its own problems. The outside of our planet is spinning much faster than the inside of it—just like a merry-go-round. Standing on a Hawaiian beach, you would be traveling 800 miles per hour faster than the core. As a result, when you jumped into your hole, you would grind against the front edge of the wall on your way down and then against the back wall on your way up.

At slow speeds this grinding would leave no more than a light road rash, but at higher speeds a continuous road rash in free fall would sand down your skin and bones until you were nothing more than falling pulp.

The smartest way to avoid the sanded-to-death problem is to begin your dig at one of the poles, where the surface of the planet is spinning at the same speed as the core.

That's step one, but death by road rash is not the only reason jumping into a hole through Earth is risky.

The terminal velocity of a falling human body in pike position at sea level is roughly 200 miles per hour. At that speed it would take 40 hours to fall 8,000 miles. In other words, depending on connections, you could book a flight to Botswana and beat yourself there the regular way. But let's assume you're not in a hurry and 40 hours is okay. You still would not make it all the way through.

Within a few seconds you would start slowing down, for two reasons.

First, as you came closer to the center of the planet there would be less Earth to pull you down, meaning you would actually start weighing less and falling more slowly. But the second, more dangerous, issue is the thickness of the air.

Mount Everest is the highest point on Earth at 29,000 feet. At that elevation there is less atmosphere above you compressing the air, and as a result the surface air is so thin, only the well trained can survive in it.

The opposite effect happens when you go in the opposite direction.

As more atmosphere is added above you, it compresses the air you're falling into. After falling only 60 miles—less than 1 percent of the way—the air would be as dense as water. For a while you would sink, but eventually you would reach a point of equilibrium where the air would be as dense as you were. So there you would stay, "floating" inside Earth for all eternity.*

Obviously, we need to make a design change to your sand pit. The solution to the air density problem is to suck all the air out of your tunnel, seal it off, and make it a long vacuum tube. That solves both the floating problem and the slow travel time, because now you would be screaming past the center of Earth at 18,000 miles per hour instead of getting stuck only partway.

Unfortunately, your tunnel still wouldn't be safe to use, because as the Russians proved when they dug the world's greatest sand hole, the inside of Earth has a heat problem.

The Russian sand pit is called the Kola Superdeep Borehole. It's the result of a massive twenty-two-year project that began in 1970 for no other reason than to see how deep they

---

*Because the atmospheric pressure would be squeezing your airspaces down, you would be denser inside Earth than you are right now, and so would sink farther than you might expect. You still wouldn't make it to the other side, though.

could dig. The Soviets made it 40,000 feet in 1989 before extreme heat melted the soldering on their drill and shut the project down. Even though they dug through less than 0.1 percent of the planet, the temperature reached 356 degrees.

The rule of thumb is that for every 100 feet below the surface you dig, the Earth heats up 1 degree, which means that after falling for 2 seconds you will be roughly one degree warmer. Not a big deal, but in your new vacuum tube you would be accelerating rapidly.

After 3 seconds your tunnel would be 3 degrees warmer, and after 30 seconds it would be as hot as an oven. It would not be comfortable, but you would survive for a surprisingly long time. In the eighteenth century the Englishman Sir Charles Blagden heated a room up to 221 degrees, sat in it for 15 minutes, and walked out unharmed. But Sir Blagden wasn't in a room that kept getting warmer, unlike your tunnel. After 30 seconds you might still be alive, but the hole would continue to heat up. After another 30 seconds you would have gone 13 miles and the temperature would have reached 1,000 degrees. If you brought a take-and-bake pizza with you, it would be ready to eat, and so would you.

But it gets worse. Not even your body would make it to the other side.

The center of the Earth reaches 11,000 degrees, hotter than the surface of the sun. At that temperature your body would instantly vaporize, which means your electrons would be ripped from your atoms and the only thing left of you would be falling bits of plasma.

So we need to make another design change on your tunnel.

We need to insulate it very, *very* (impossibly) well. Would you make it?

Assuming you didn't hit the sides of your tunnel—which would slow you down and leave you short of the other side—you would reach the center of the Earth in just over 19 minutes and be falling at 18,000 miles per hour. Once you passed the center you would begin slowing as more and more planet began pulling you back. But, just like on a playground swing, your momentum would carry you back to the same height at which you started—in this case the other side of the Earth.

So if you ignore the impossibility (with current technology) of digging a hole in the extreme temperatures and pressures of the Earth's core, would you make it to the other side? Actually, yes! Approximately 38 minutes and 11 seconds later you would reach the other side of the globe. Just make sure to grab the surface when you get there.

Miss, and you would start the whole process over again.

What Would Happen If...
You Toured the Pringles Factory and Fell off the Catwalk?

YOU HAVE PROBABLY taken a factory tour at some point in your life. Not terribly exciting, but maybe that's because you didn't become a part of the product. Let's change that.

Let's say you were walking above the factory floor at the Pringles potato chip factory and just as you were admiring a shipment of potatoes, you fell into them.

As far as we can tell no one has ever died in the Pringles plant, but you would hardly be the first to die in an American factory.

In the five-year period between 1902 and 1907, for example, more than five hundred American workers died every year in factories. *The Factory Inspector*'s annual report chronicled some of the accidents in its yearly roundup:

A worker in a brick-making factory was caught in a belt and had most of his skin torn off.

A sawmill worker fell onto a large, unguarded circular saw and was split in two.

A worker got caught in the large flywheel of the main steam power plant of a navy yard, his arms and legs were torn off, and his lifeless trunk was hurled against a wall fifty feet away.

And on it goes. Pringles chips weren't invented until 1967, long after factory safety standards had improved, so no one's been turned into a Pringles chip yet. But if you fell into the potatoes, you would change that. Here's how it would go.

Once you were in with the raw potatoes you would head to the first stop: the heater.

To make a chip, potatoes are first dehydrated for consistency and preservation with a blast of 600-degree air. While humans are better at retaining water than potatoes are and you would not be completely dehydrated, your cells would not like the extreme heat.*

Human cells are capable of functioning at a body temperature of up to 113 degrees, but a 108-degree fever is often fatal because your cells have a self-destruct button as an adaptation to fend off disease.

When a virus infects your body, it commandeers your

---

*A better way to dehydrate a human is the freeze-dry method, where a person is frozen solid and then allowed to dry in an arid environment. The five-thousand-year-old Ötzi iceman is an example of this occurring naturally. A glacier covered Ötzi shortly after his death and his body was so perfectly preserved, scientists were able to determine how he died (murder—an arrow severed the artery in his shoulder), what he ate as his last meal (grain, roots, and fruit), and perform a blood analysis (he was lactose intolerant).

cells and turns them into little virus-making factories. Infected cells break open and release the viruses to infect other cells. To slow a virus's growth, your cells interpret a high body temperature as a signal that you're battling a virus and destroy themselves before they're commandeered, like a self-destructing message in *Mission: Impossible*.

Back at the factory, your cells would misinterpret your body temperature rising from the heat of the oven as a fever and begin to self-destruct. At 108 degrees you would have lost so many brain cells that you would lose control of critical functions like your heartbeat.

From there you would be mashed and minced into a fine powder. Then some corn and wheat would be mixed in with your powdered body to create a formula resembling pancake mix. After that, water would be added until you were a nice slurry, and from there you would go to a rolling press that would flatten you with four tons of pressure.

If you stuck your hand into the press it would be flattened into the size of a basketball. But fortunately you're already dead and are now a powdered slurry, so all it would do is stamp your human pancake mix flat.

Next, your thin sheet of body would be cut up into small chip-size ovals, with the remainder peeled off and recycled back into the process. From there you would be molded into the familiar concave chip.

Your new shape—called a hyperbolic paraboloid—was not created haphazardly, by the way. Its design was one of the first commercial uses of a supercomputer. Your new form is perfectly *un*aerodynamic so that it won't fly off the factory conveyor belt and crafted so your pieces fit snugly into the can.

Once you're in the familiar Pringles shape you would be immersed into a deep fryer for exactly eleven seconds. At this point you would be dead by way of heat, powdering, pressing, and cutting.

After that your fried remains would be lightly flavored. In the United States the flavors are usually salt and pepper, or maybe ranch. If you wanted to be turned into something more interesting, you should have fallen into the Pringles factory in Mechelen, Belgium, where they produce flavors like wasabi and prawn cocktail.

Your final flavored parts would then be stacked and placed into Pringles cans. At this point, you would be the first human Pringles chip, but, interestingly, you would not be the first human buried in a Pringles can. That distinction belongs to Fred Baur, the inventor of the Pringles can, who requested that his ashes be placed in his invention.

You would, however, be the first ever placed into *multiple* Pringles cans. Let's assume you fell into the potatoes weighing 180 pounds. Once the water is removed from you, you would lose 60 percent of your weight, but then because only 42 percent of Pringles is potato you would gain much of that weight back in both corn and wheat. In the end, after a bit of back-of-the-envelope math, we think you would be processed into roughly 40,000 Pringles chips, thereby filling just over 400 cans. In a typical day Americans consume 300 million Pringles chips—that's 3 million cans—so the odds that any single potato chip consumer would enjoy all of your ranch-flavored remains are quite low. But a few unlucky souls would get a full can of you, and, thanks to a rather gruesome experiment run by the American journalist William Seabrook in the early 1900s, we have some idea of what those unfortunate few would taste.

With the help of a hospital, Seabrook acquired a chunk of human meat from a recently deceased person, and after cooking and preparing it he reported that "in color, texture, smell as well as taste . . . veal is the one meat to which this meat is accurately comparable."

Exactly how a chip composed of 42 percent veal, some corn, wheat, and ranch seasoning tastes we will leave to an adventurous reader.

What Would Happen If . . .
You Played Russian Roulette with a Really, Really Big Gun?

QUESTION: IF YOU played Russian roulette with a million-chamber gun, would it add any meaningful danger to your life?

Answer: If your only activity on a single day was to hoist a million-chamber gun to your head and pull the trigger one fateful time, it would be the safest day of your life.*

All the basic risks you take every day—walking a few blocks, driving a couple of miles, walking underneath air conditioners—taken together add up to about 1.5 times the danger of playing a single round of enormous-gun Russian roulette.

Ronald Howard, a professor of decision analysis at Stanford University, needed a way to measure the tiny risks of everyday activities against one another, so he coined the term

---

*This ignores the risk of your dropping the gun and crushing yourself. A million-chamber Smith & Wesson would weigh around 250,000 pounds. A bullet to the head would be the least of your concerns.

micromort—the one-in-a-million probability that a given activity will kill you.*

You can use micromorts to measure the risks of different modes of travel. Driving 250 miles in a car equals a single micromort. Motorcycling—or canoeing!—for only 6 miles is equal to the same. Flying in a private aircraft is only slightly safer at 8 miles per micromort. Walking (17 miles) and bicycling (20 miles) are safer still, but by far the safest modes are commercial flying (1,000 miles) and riding on a train (6,000 miles).

If you're adventurous, the idea of playing this version of roulette should seem tame. Going for a swim in the ocean? That's 3.5 micromorts. Scuba diving? 5 micromorts per dive. Running a marathon is a surprisingly high 7 micromorts per run.† White-water rafting? 8.6 micromorts per day on the river. Skydiving goes up to 9. The average adventurer seems willing to risk 10 micromorts for a thrilling experience, but true daredevils risk far more.

A BASE jumper, for example, risks 430 micromorts per jump. A climber that goes beyond base camp at Mount Everest risks 12,000 micromorts (a 1-in-83 shot at death). And for

---

Micromort is a conjunction of "microprobability" (the one-in-a-million chance of something occurring) and "mortality" (your dying).

†The most common cause of death from running is heart attack, usually the result of an underlying heart condition. Another issue is an otherwise rare condition called hyponatremia. When your body sweats you lose not only water but salt. If you replace the water but not the salt, sodium levels in your blood drop and water rushes into your brain cells and inflates your brain. Not good. As it swells it presses against your skull, causing nausea and short-term memory loss, and it is fatal if left untreated.

every ten people who have reached the top of the mountain K2, three have died.

For those of us who don't BASE jump or climb Himalayan mountains and are younger than eighty, our very first day is probably the most dangerous of our lives. At 480 micromorts, that day is the equivalent of a cross-country motorcycle trip.

We also put a dollar value on our micromorts, whether we think about it consciously or not, and are willing to pay to reduce them. To mitigate everyday risks, the average American will spend fifty dollars on extra safety features, like adding optional airbags, to avoid 1 micromort. However, your government doesn't value your micromorts quite as much as you do. When deciding whether to make road safety improvements, the Department of Transportation looks at how many micromorts the improvement is expected to save and divides that by the cost. If the price for saving each driver 1 micromort is more than the price of a Big Mac, they don't make the change.

There are losers to this game, though, which brings us back to our original question regarding Russian roulette with the million-chamber gun. For every million people who play there will be, on average, one person who runs out of luck.

But wait! Just because you have shot yourself in the head doesn't mean you're going to die. It just means you're probably going to die. Of victims who are shot in the head, 5 percent survive the injury. The reason? Redundancy. The brain can transfer jobs from one hemisphere to the other, and essential functions are done in both hemispheres. The brain's hemispheres are divided left to right, and a bullet that destroys only one hemisphere or part of one hemisphere gives you a better chance of survival—meaning that a bullet that enters your forehead and leaves out the back of your head is slightly more survivable than one that goes from ear to ear (see p. 34 for how you could live if you don't put just a bullet through your head but an entire rod).

The speed of the bullet to your head is also important. A high-speed rifle shot can hit your skull and skip in unpredictable ways, just like a stone skipping across water. This means a direct shot to the forehead might hit your skull, skip upward, and miss most of your brain.

A bullet from a handgun would hit your skull and travel straight like a slow-moving stone. That's bad if its aim were true.

Even a handgun's bullet travels faster than your tissue can tear, meaning it would push your brain out of the way as it traveled. If you could take an X-ray while the bullet was still in your head, you would see a wake behind the path of the slug wider than the bullet itself.

That X-ray would hide what was actually going on, though. Not only would the tissues and nerves in the bullet's path be destroyed, but so would a large area on either side of it.

As the bullet passed through your brain matter the tissue would collapse back together—like water smashing together in the wake after a dive. This cavitation in your brain happens quickly, and the tissue would collapse with enough force to send a shock wave that would destroy your nerves in a wide swath.

If you were to survive the immediate shot, the areas that were damaged would dictate what type of recovery you would make. But because of the brain's ability to transfer jobs, it's impossible to predict exactly how you would recover.

In nearly every case, the first thing a person thinks after a bullet to the head is that something is burning. For reasons not entirely known, brain damage often makes victims smell burned toast.

In all likelihood, though, you would not have to worry about that. A point-blank shot would probably kill you before your brain was able to process what happened.

In other words, after being extremely unlucky to lose your one-in-a-million bet, you would be very lucky to survive it.

What Would Happen If . . .
You Traveled to Jupiter?

Aт 3:21 EASTERN Standard Time on October 9, 2013, NASA's *Juno* spacecraft whipped by Earth at 25 miles per second—that's fifty times the speed of a bullet—and raced toward Jupiter on a mission to collect data. The probe was unmanned, but let's say you jumped on, finally arrived at Jupiter in July 2016, and jumped out. Here's what would happen.

Jupiter is a gas giant, so parachuting through it might seem as harmless as passing through a cloud. That is not the case. Jupiter's mass is enormous, but its heat is intense and there's enough pressure inside it to put our deepest oceans to shame. The planet is so impenetrable that we're not even sure what makes up its core. So far it has gobbled up our probes long before they could get more than a few miles below Jupiter's cloud tops. In 1995 the *Galileo* orbiter dropped a probe into Jupiter. It managed to transmit for fifty-eight minutes before it was crushed and incinerated. You would not be so lucky.

The trouble would begin long before your jump.

Jupiter's magnetic field stores the sun's radiation like a battery, the same as Earth. But Jupiter is bigger than Earth and its magnetic field is far stronger, so even 200,000 miles away from Jupiter you would be zapped by 5 sieverts (Sv) of radiation, which is enough to kill you after a few days of exposure. And as you got closer to the planet the dosage would increase to 36 Sv (10 Sv is a lethal dose), which would immediately induce vomiting and eventually death.

But let's say you came prepared for this with a radiation shield on your space suit—lead and paraffin wax would work—and you survived long enough to jump.

Once your feet left the probe's deck, Jupiter's huge gravitational pull would accelerate you to more than 30 miles per second—for comparison, a .50-caliber bullet travels at a relatively pedestrian 0.5 miles per second.* When you entered Jupiter's upper atmosphere you would begin to slow down from 30 miles per second to 4 miles per hour in less than 4 minutes. At the peak of your deceleration you would experience 230 g's—the equivalent of face-planting off a sixteen-story building.

In addition, falling at 30 miles per second means the air cannot get out of the way fast enough, so it's compressed and superheated. Your space suit would warm up to more than 15,500 degrees, vaporize, and turn you into a ball of plasma while producing a light brighter than the sun.

---

*This acceleration would kill you if you did it in a rocket ship because the back of your seat would push its way through your organs. But in a space suit on Jupiter you would be fine (for now) because when gravity is doing the accelerating, everything in your body accelerates at the same speed, so no organ pileup.

From the surface—if there was one and if there was anyone to look up and see you—you would look like a streaking
ball of fire. But the *Galileo* probe was able to survive this process with a sophisticated heat shield that ablated and blew
away as it entered the atmosphere, so let's say you grabbed
one of those before you left and survived your entry.

At this point we can say you have reached the surface of
Jupiter, only what looks like the surface is just the cloud tops.
Since Jupiter is made of gas, you would continue to fall. At
1 atmosphere of pressure here on Earth, a human body's terminal velocity is about 200 miles per hour in the pike position. But Jupiter's gravity is much stronger than Earth's. At
1 atmosphere on Jupiter you would fall at 1,000 miles per
hour—still fast, but at least you would have slowed enough so
that your suit would no longer be melting. The temperature
outside would be a chilly 135 degrees below zero with an atmosphere of mostly hydrogen and helium, but in your suit
with an oxygen tank and a heater you would be okay.

After ten minutes of continued falling you would reach
3 atmospheres of pressure, or the equivalent of being 100 feet
under water. Fortunately your body is mostly water, and water is incompressible. Professional free divers can drop 700
feet in under three minutes, where the pressure is 21 atmospheres. Not terribly safe, but survivable.

As you get closer to the core, Jupiter's temperature rises, just
like Earth's, and by this time the temperature would have risen
to 100 degrees below zero. The clouds are made of ice particles—
similar to the upper atmosphere of Earth—and winds would

have picked up to 450 miles per hour. But, assuming you made it this far, you would probably be okay inside your space suit.

After twenty-five minutes of falling, the temperature would rise to a balmy 70 degrees. The pressure would increase to 10 atmospheres—the equivalent of 330 feet of water. At 10 atmospheres of pressure, oxygen becomes toxic. To stay alive, you would need to switch tanks to one with a helium-oxygen mix like the kind deep-water scuba divers use.

After a full hour of falling you would be in real trouble. It would be completely dark outside and the temperature would have reached 400 degrees—hot enough to kill you within a few minutes and to melt the solder on the *Galileo* probe. Your only saving grace at this point would be if you were in a well-insulated space suit. Let's say you were.

The atmosphere would continue to increase in density as you fell, becoming as dense as water, then as dense as rock. You would never encounter a surface on Jupiter—the atmosphere just smoothly grades denser and denser in the increasing pressure.

Eventually your density would reach an equilibrium with the planet's, so you would stop sinking and hover in place. The pressure would now have increased to 1,000 times Earth's atmosphere. Even your special space suit couldn't withstand that. It would collapse along with the air cavities in your body. Your chest, ears, face, and gut would fall in on themselves until you were a solid mass of flesh and blood. And then there's the heat.

The temperature would be 8,500 degrees at this depth—about the same as the surface of the sun. Not only would you vaporize but your atoms would break apart. You would become permanently entombed as bits of plasma wafting in the pitch-black, searing-hot liquid hydrogen.

If you managed to make it deeper into Jupiter the pressure would eventually be more than 1 million atmospheres and something interesting would happen: 62 percent of the atoms in your body are hydrogen, and at that pressure scientists predict that hydrogen turns into a liquid metal.

So if you somehow made it past the g-forces, heat, pressure, and poisonous atmosphere, you might resemble the bad guy from *Terminator 2*. That's cool, at least.

What Would Happen If...
You Ate the World's Deadliest Substances?

On November 1, 2006, Alexander Litvinenko sat down to eat with two former KGB officers in London. Litvinenko was a former Russian security officer who publicly opposed the current regime, worked for the British secret service, and wrote articles accusing Russian president Vladimir Putin of terrorist acts and assassinations.

Soon after the meal Litvinenko felt ill. At first the symptoms resembled food poisoning: vomiting, upset stomach, and fatigue. But, unlike food poisoning, the symptoms intensified over the following days and doctors could find no explanation. Litvinenko's hair fell out, his blood cell count plummeted, and eventually he couldn't get out of bed. He died three weeks later.

Through an autopsy, investigators determined that Litvinenko was poisoned with 10 micrograms (half the weight of an eyelash) of polonium-210, a toxic radioactive isotope that occurs as uranium decays into lead.

Polonium-210 has a short half-life—only 138 days—and in that time releases a huge amount of energy. One gram will heat up to 900 degrees and generate 140 watts of power. It's used on spacecraft as a heat and power source, and would make for the world's greatest ski boot and glove warmer.

Polonium-210 is so reactive and its alpha radiation so terribly destructive, it dissipates all its energy over a very short distance, which means it can be blocked by clothing, two pieces of paper, or even skin. Litvinenko's killers could have easily carried it in a pocket, probably in a vial of water, and been fine.

Once given an avenue past your skin, however, such as ingestion, polonium-210 becomes incredibly toxic, and death from radiation poisoning is inevitable. It doesn't make a great weapon for an assassin, however, because it can be traced in a way that puts even the greatest bloodhound to shame. Apparently the ex-KGB officers weren't aware that equipment exists to detect it in fantastically small quantities, and investigators followed the killer's trail from his contaminated aircraft to all three of his hotels to his rendezvous with Litvinenko, and to Litvinenko's teacup. (The Russian government declined to extradite the accused.)

As soon as Litvinenko drank his poisoned tea he was doomed. Once given access past the skin, polonium-210's alpha radiation begins its bombardment of the body, starting with the lining of the stomach and gut, causing severe nausea, pain, and internal bleeding. The earlier these symptoms present themselves, the higher the dosage you received. If they begin within four hours after exposure, you're in trouble.

Your bone marrow, which directs blood production, is particularly susceptible to radiation. As those cells are attacked and destroyed, your white and red blood cell counts fall and you become vulnerable to outside infection.

As more bone marrow is destroyed, fewer red blood cells are created. Eventually the blood becomes so thin you're unable to oxygenate vital organs—the most important of which is your heart. Once the heart stops receiving enough oxygen it will fail and cut off all blood flow to the brain.

Polonium-210 has a lethal dose of a single microgram, which makes it the deadliest radioactive material, but it isn't the deadliest substance in the world.

As bad as polonium is, botulism is five hundred times more toxic.

In 2013, the California Department of Public Health received a stool sample from a baby suffering from botulism. Babies have an undeveloped gut and will occasionally develop botulism when an adult would be able to fend it off.

The test is fairly routine, and with an antibotulism serum the survival rate is good. This time, however, the doctors discovered something different. They called it botulism H, a previously unknown type of botulism that's unfathomably toxic with no known antiserum. The discovery so alarmed researchers, they have kept the DNA sequence a secret to prevent production and weaponization.

Botulism H is fatal at 2 *nanograms*. That's 2 billionths of a gram. A single red blood cell, which is completely invisible to the naked eye, weighs 10 nanograms. The deadliest chemical weapon ever created, VX gas, is nasty stuff but requires a dose of 10 milligrams to kill you.* That makes it more than a million times less potent.

Here's how toxic the botulism H toxin is: If you put it into an eye dropper and squeezed a single drop into a swimming pool, drinking a glass of water out of the pool would be fatal. That same drop, properly dispersed, could kill a million people. A cupful could wipe out Europe.

Unlike a virus, botulism H doesn't grow once it takes up residence in your body—another remarkable aspect of the toxin. It starts very small, stays very small, and still grinds your body's functions to a halt.

Muscles contract as a reaction to a chemical called acetylcholine. Botulism slides into your muscles' acetylcholine receptors and takes up permanent residence, effectively paralyzing you.

This feature actually has a number of medical uses. A different strand, botulism A, is used in cosmetics. An injection

---

*A quick primer on VX: It was developed as a pesticide until people realized it was far too toxic for that. The military took notice of its toxicity, however, and turned it into a chemical weapon.

Here's how it works: Your nerves spit out chemicals that tell your muscles to contract and relax. VX gas disables the "relax" chemical, so your muscles clench up but don't unclench. Unable to relax, your muscles quickly tire and stop working. This is a problem, particularly for your diaphragm. Once exposed to VX your diaphragm seizes up, tires, and you die of asphyxiation. The entire process takes only a few minutes.

Unlike in *The Rock*, VX gas does not have any effect on your skin and the antidote is injected into your thigh, not your heart.

of a tiny, *tiny* amount of it can relax the muscles in your face and eliminate wrinkles. Its commercial name is Botox.

But there is no commercial application for botulism H.

If you took a drink from that contaminated pool, twelve to thirty-six hours later your vision would get a little blurry, your eyelids would droop, and your speech would slur.

Botulism attacks the muscles controlled by your cranial nerves first—your eyes, mouth, and throat—and from there it spreads. Constipation is next, after the muscles that keep your meals moving are paralyzed.

One of the scarier aspects of botulism poisoning is that it has no effect on your mental state. As the wave of paralysis passed down your body you would be completely aware of what was happening, but neither you nor your doctors would be able to do anything about it.*

It would begin at your head. After your face froze, your shoulders and arms would follow suit.

The trouble begins once your diaphragm stops working. The muscles in your chest allow your lungs to expand and fill with air. As they became paralyzed, you would struggle more and more just to breathe, as if a five-hundred-pound man were sitting on your chest.

Eventually you would no longer be able to get enough air to sustain your brain. Brain cells need a constant supply of oxygen and start to drop off after only fifteen seconds of deprivation. A

---

*Patients of the more common forms of botulism—for which there exists an antiserum—can lie in bed for months paralyzed from head to toe but with fully functional minds. The antiserum stops the progression of botulinum toxin, but nerves already blocked are dead forever and patients have to wait months or years to grow new ones.

few minutes later—the timing depends on the order in which your brain cells died—you would suffer complete brain death brought on by a dose of poison smaller than the period at the end of this sentence.

On the positive side, your corpse would be smooth and wrinkle free.

What Would Happen If . . .
You Lived in a Nuclear Winter?

DURING THE COLD War it was widely understood that both the United States and the USSR had the capability to destroy the world with nuclear weapons. What people didn't know was how easily they actually could do it.

Today, thanks to sophisticated weather models built to analyze global warming, we know that even a relatively small nuclear skirmish would be extremely bad news. Simulations of full-scale war between smaller nuclear-armed countries suggest that a hundred multimegaton bombs would be exchanged, and the simultaneous detonation of a hundred nuclear devices would be bad for you even if you were on the other side of the globe. Your first problem? Radiation.

When the nukes went off they would irradiate the area and transmute innocuous atoms into dangerous ones. One of the worst of these nuclear bastard children is called strontium-90.

It's light, so it doesn't take many explosions for it to coat the globe and get deep into the food supply. Once ingested it's so similar to calcium that your body absorbs it into your bones. Children born after the open-air nuclear tests of the 1950s have fifty times the natural level of strontium-90 in their teeth. Fortunately, that's still below the threshold for serious danger. Unfortunately, unlike a test, a nuclear battle will blow past that threshold.

Once strontium-90 is in your bones its radioactive decay breaks up the DNA of your cells, leading to bone cancers and leukemia. So if you survived the initial nuclear exchange, you would have bone cancer to look forward to, but that's only if you could also survive the more serious smoke, ash, and soot problem.

The second issue, after the dust has cleared from the initial detonations, is that the dust *wouldn't* clear. After a hundred multimegaton bombs exploded in the air, not only would they directly distribute carbon into the upper atmosphere but they would start enormous forest and urban fires that would release massive amounts of smoke. On top of that, the explosions would lift tons of fine dust—all of which would be heated by the sun to rise and collect in the stratosphere.

The smoke from your typical campfire stays below the clouds where it can be wicked away by rain. In the case of nuclear fallout, smoke and ash would be lifted above the clouds where it wouldn't be wicked away by rain, so it would stay parked for years and block sunlight.

Even conservative environmental simulations show that a hundred nuclear detonations would block enough sunlight to drop the average global temperature by a few degrees. A sudden global drop in temperature of even a couple of degrees would be devastating for the world's food supply, because a single frost kills rice. A serious disturbance in rice production would kill as many as 2 billion people around the globe.*

In a hundred-bomb nuclear war nearly a third of the world's population would die from the explosions, starvation, or cancer, but our species would carry on. In larger, multithousand exchanges of thermonuclear weapons, like the one that almost occurred in November 1983 between the United States and the USSR, however, we probably wouldn't.

On November 7, 1983, the United States led NATO in a massive training exercise called Able Archer that mimicked a nuclear first strike against the USSR. Unfortunately for nearly everybody, the USSR believed the exercise was a cover for an *actual* first strike. So in response the Soviets helicoptered missiles to their silos and mobilized their air force, actions that should have alarmed the U.S. military to respond in kind. Luckily, Lt. Gen. Leonard Perroots of the U.S. Air Force mistook Russia's actions for a simple training exercise and took no action. That lack of response convinced the USSR to stand down.

Lt. Gen. Perroots made a "fortuitous, if ill-informed" deci-

---

*Global rice production would drop 21 percent, corn output would decrease by 10 percent, and soybeans would decrease by 7 percent, according to an analysis by the International Physicians for the Prevention of Nuclear War.

sion, according to declassified analysis of the scare. It might be the most fortuitous mistake in human history.

If the alarm had been raised and the misunderstanding escalated into full-scale nuclear war, a few thousand multi-megaton bombs would have crisscrossed the globe and deto-nated over their targets. Even if you didn't live in a large city (basically every city with a population of more than 100,000 in the United States and the USSR was targeted) and there-fore weren't killed by the initial blasts, you couldn't expect to live long.

Within two weeks of something like this happening, 180 million tons of smoke, soot, and dust would coat our globe like black paint, and there it would stay.

Light levels would be reduced to a few percent of what they are today, so high noon would look like predawn. Midsum-mer highs in North America would be below zero.

The good news is there would be plenty of dead trees to burn for warmth. The bad news: You would starve. Crops would be wiped out, and those that weren't would suffer from another problem: bugs.

Cockroaches and their ilk are quite durable when it comes to radiation, but their predators are not. Without any birds to keep them in check, crop-eating pests would pros-per. Pests would decimate any crops that made it through the freeze.

But there is an upside (sort of). Cockroaches are actually more efficient than cows at turning grain into protein, and even in the new apocalyptic world there would be plenty for them to eat. They're also a healthy snack. Cockroaches are high in vitamin C, protein, and fat, so as long as you're not a picky eater, you may survive a little longer than expected.

You would just need to eat *a lot* of cockroaches, around 144 per day, to survive. Gross.

What Would Happen If...
You Vacationed on Venus?

VISITING VENUS DOES not offer the cornucopia of death that parachuting onto Jupiter does, but it's still no picnic.

The descent into the Venusian atmosphere from deep space should be relatively pleasant. Its gravity is similar to Earth's so you would not fall too quickly—it's roughly the same as Earth's reentry, reducing this to an already solved problem. All we have to do is put you in a NASA space shuttle and you would arrive 155,000 feet above the planet in one piece (if you decide to skip the shuttle, refer to p. 77 to see what would happen).

However, once you descended below that 155,000-foot mark, your troubles would begin.

First, you would need to watch out for the rain clouds, because on Venus it doesn't rain water, it rains sulfuric acid—similar to what you find in car batteries. The rain would eat away at the exposed metal of your shuttle (and if you didn't have a shuttle it would bore holes in your skin). A window in your shuttle would have to be made of diamond, which makes for a fantastic option due to its resistance to heat and sulfuric acid. The NASA Venus lander used a 205-carat industrial diamond as a lens for its camera.*

The second reason storm clouds could be dangerous is lightning. Scientists only recently confirmed lightning's existence on Venus, but still do not know for sure whether it's only intracloud lightning or whether it strikes the planet. Either way, if you're inside your space shuttle it will conduct the electricity around you and protect you just like a car does on Earth. If you're outside the shuttle and you're hit by a sulfuric-inspired bolt, refer to p. 67 to see what would happen. It's not pretty.

Once you descended below the clouds you would need to slow yourself down with a parachute. Unfortunately, Venus has a greenhouse-gas problem—a really, really big one. Its atmosphere is 96 percent carbon dioxide, which means it's a phenomenal heat trap. The planet's temperature is 864 degrees during the day and it's so good at storing heat that it's still hot enough to melt lead at midnight. Think global warming to the max.

Standard polyester or nylon parachutes melt at 270 degrees. Your parachute would be gone seconds after it deployed. We recommend using Dacron, which is what the Venus lander used and is both resistant to sulfuric acid and melts at

---

*If you can prove you have scientific intentions, the government will give you confiscated diamonds.

500 degrees. So it would still melt, but at least it would hang in there for a little while, and that actually might be enough because the air is so thick on Venus—7 percent the density of water—your crash landing would be slow enough for you to survive.

When the Russians put a lander on Venus they used a combination of melting parachute, puffy balloon, and crash landing—and together they worked. The lander broadcast fifty-two minutes of data before the heat melted its electronics.

So with a bit of luck and engineering (not to mention an unbelievable air-conditioning unit) you might get your feet on Venus long enough to have a look around. You would probably be disappointed. The planet is perpetually covered in a seventeen-mile blanket of smoglike clouds that would make Los Angeles look like a Tahitian island. The smog is so thick that high noon on Venus would look like dusk.

The gravity is 90 percent of Earth's, so your body can easily adapt, but the "air" is fifty times denser than Earth's, so running would be in slow motion, like escaping an ax murderer in a dream.

The denser atmosphere also causes a problem for the air cavities in your body. Standing on the surface of Venus is the equivalent to being three thousand feet under water. Most of your body is made of water and therefore incompressible, but you do have a few air cavities in you. Those areas collapse under pressure. Your face would smash in on itself as if it were hit by a huge bat; your ears would crunch inward and your eyeballs would fall into your head. You would drop a few collar sizes as your throat and larynx cinched shut, and you would lose a few inches off your belly when your stomach and intestines collapsed inward.

Your lungs are the largest gaseous areas in your body and would also collapse, but they would be useless on Venus anyway, even if you could somehow keep them inflated. Since Venus's atmosphere is 95 percent carbon dioxide, after a single breath your body would absolutely scream for oxygen. The fifteen seconds before you went unconscious would pass painfully.

The final issue on Venus is, of course, the heat. If you were in a swimsuit at 870 degrees, you would be dead in seconds, although you would not burn because there is no oxygen to support combustion. Even though you would not ignite, your cells stop working in 870-degree heat, more or less the temperature of a well-stoked fire, and your proteins denature. You would quickly progress from "well done" to "smoldering bones" and, eventually, "ash" over the course of a few days.

There are many ways to die on Venus: crematorium-level heat; crushing, deep-ocean-level pressure; a general lack of anything to breathe.

However, there is one way in which you would absolutely not die: falling.

The air is so thick your terminal velocity on the planet would be eleven miles per hour—the same speed you reach when jumping off a five-foot ledge here on Earth. Meaning no matter how high a cliff you jumped off on Venus, and although you would probably die *while* falling from a few different means, you could never fall to your death.

In summary, Venus is a terrible place to visit if you don't want to die in a furnace, but a fantastic place to go if you're afraid of heights.

What Would Happen If . . .
You Were Swarmed by Mosquitoes?

THE FEMALE ANOPHELES mosquito is the single most dangerous creature in human history. According to some estimates, its bite is responsible for half of all human deaths since the Stone Age. Of course, you should not give the mosquitoes all the credit. The real killer is malaria, a disease caused by a parasitic protozoa that hitches a ride on mosquitoes.

More than 247 million people are infected with malaria every year. More than 1 million die from it. On top of that a mosquito bite is annoying (their saliva is an anticoagulant, which most of us are allergic to), and we're not the only ones who think so. The Alaskan caribou alter their migration routes into colder areas to avoid the bites.

Of course, caribou aren't the only ones that avoid areas because of mosquito infestation. Huge swaths of jungle in Central America, South America, and Africa were completely impenetrable to early explorers thanks to the mosquito. The preservation of the Amazon rain forest can largely be attributed to it.

The first attempt to build the Panama Canal was a French-led effort that began in 1881. It did not go well. The Panamanian jungle was filled with poisonous snakes and spiders, which didn't help, but their danger paled in comparison to the mosquito problem. Malaria absolutely decimated the French workforce. Mosquitoes killed nearly 200 workers per month at the height of the project. Deadlines were missed, costs ballooned, and after nine years the project ended in failure. In all, 22,000 workers died—nearly all killed by the mosquito. It wasn't until the U.S. project twenty years later—and after doctors better understood the link between malaria and the mosquito—that the canal was completed, and still at the cost of 5,600 more lives.

There still remains a question for those of us who swat a lot of mosquitoes but don't live in malaria country. Can mosquitoes kill you *without* the help of the protozoa? Can enough mosquito bites suck you dry? Is there such a thing as death by a thousand bites? Mosquitoes take a small amount of blood each time they bite, which isn't a problem on your typical camping trip. You can afford to lose it. But it *could* become a problem if you happen to be camping on Alaska's North Slope and find yourself naked in a large swarm. We know the specifics of this fate thanks to researchers in the Arctic who, after what we assume were a lot of dares and at least some vodka, ventured outdoors without shirts on. For one minute they stood outside in the thick cloud of mosquitoes before scampering back inside and assessing the damage.

They each counted more than 9,000 bites.

Mosquitoes drain only 5 microliters of your blood per bite, and you have roughly 5 liters pumping through your veins, which equates to roughly 1 million mosquito meals. So you can afford a few bites on your next camping trip, but 9,000 per minute is a different story.

If you followed in those brave shirtless scientists' steps, but then stayed in the swarm, here's what would happen.

Roughly fifteen minutes into your ordeal you would lose 15 percent of your blood, which is about the same amount taken at a blood bank. You would experience some slight anxiety and a lot of itchiness, but nothing a glass of orange juice and a cookie couldn't fix.

After just over thirty minutes, however, the mosquitoes would have sucked 30 percent of your total volume. Your blood pressure would begin to drop and your heart would be forced to speed up to compensate. At the same time you would begin to feel a cooling in your extremities as your body focused on providing oxygen to your internal organs at the expense of your hands and feet. Meanwhile, your breathing rate would increase as your body tried to compensate for the oxygen deficit.

Forty minutes into the biting you would have lost two liters and reached a critical stage. You would be anxious and confused, with your heart racing at more than one hundred beats per minute. As your body concentrated the remaining blood and oxygen to your brain, kidneys, and heart, the tissue in your arms and legs would begin to starve and die.

After forty-five minutes and more than 400,000 bites,

you would have lost more than 2 liters of blood. At that point your heart would no longer be able to maintain the minimum necessary blood pressure and you would go into shock followed by cardiac arrest. Without the blood flow needed to carry oxygen from your lungs, your brain cells would begin dying. Within a few seconds you would enter an unconscious state and suffer irreparable brain damage. Depending on which brain cells died and in what order, you would have between three and seven minutes from heart failure to total brain death.

And, in a most unusual way, you would join nearly half of all mankind to succumb to the mosquito.

What Would Happen If . . .
You Became an *Actual* Human Cannonball?

Human cannonballs—the kind that you see in the circus—are shot from circus cannons (basically long tubes with springs in the bottom). The record ride is about 200 feet, which, if you do the math, would take something like a 70-miles-per-hour launch. With a properly placed net, the experience is survivable, although not terribly safe. More than a few have died on the job. But even so, it's a lot safer than getting shot from a real cannon.

Real cannonballs these days leave the muzzle at several thousand miles per hour. Let's say you wanted to see what that felt like and so you crawled inside a modern cannon and found a friend to shoot you out. There are many hazards associated with this idea, but we'll talk about only two.

One, the acceleration problem. The instant your friend pulled the trigger, you would go from zero to 3,800 miles per hour in about 1/100th of a second. This is the equivalent of

17,000 g's, roughly 2,000 times greater than any astronaut has ever experienced. For a moment, you would "weigh" 2.5 million pounds. Your skull and your bones would instantly collapse along with all your soft tissue (your organs, flesh, muscle, etc.). Only the water in your body would resist. So while still in the barrel, you would have lost your human form and become a small cylinder of reddish water with a very thin scum of crushed bone and flesh at the bottom of it. After you exited the barrel, it would only get worse.

Traveling at 3,800 miles per hour creates a tremendous amount of friction with the air, which results in heat (the surface of a fighter jet reaches 600 degrees). That would be a problem for the water that would now make up the majority of your corpse.

So only in your fondest dreams would you be a thin disk of reddish water flying through the air. Your final form would be a superhot mist ejected into the atmosphere at five times the speed of sound.

Ouch.

You Were Hit by a Penny Dropped from the Top of the Empire State Building?

THE BAD NEWS: A penny dropped on your head from the top of the Empire State Building will not bore a hole straight through your skull. Its terminal velocity is only 25 miles per hour at sea level. The penny is both light and, like all coins, tumbles as it falls, which adds to its surface area and makes it an especially poor lethal projectile. Not even the Eisenhower silver dollar, the largest coin in circulation, would do more than sting.

Everyone is always disappointed to learn this. The image of a smoking penny-size hole through your head is so compelling that most people aren't willing to give it up easily.

However, there are some objects that will do more damage if they're dropped from the top of the Empire State Building, but as the penny example demonstrates, it isn't always intuitive which you should try to catch and which you should run from. In response to this common urban-dweller's dilemma, we have created a guide to walking under the Empire State Building.

Here's what you should do if you see these objects falling from the top of the skyscraper.

Baseball

A five-ounce baseball dropped from the top of the Empire State Building would top out at 95 miles per hour, about the speed of a major league fastball.* If it bounced off your head you would probably suffer a concussion. But there is also an opportunity for a record here.

In 1939, Joe Sprinz, a catcher for the San Francisco Seals, set a world record by catching a baseball dropped 800 feet from a blimp. The ball hit his glove with enough force to smash it into his face, break a few teeth, and fracture his jaw.

In 2013, Zack Hample extended the record to 1,052 feet (he wore a catcher's mask). Since the Empire State Building is 1,250 feet tall, you could set a new mark. Or concuss yourself in the effort.

Conclusion: If you see a ball falling from the ESB, grab a glove—and maybe some protective gear as well. Baseballs traveling slower than 95 miles per hour have killed people.

---

*If you're comparing the speed to a major league pitch, it's actually equivalent to catching a 103-miles-per-hour fastball because radar guns measure the speed of the ball when it leaves the pitcher's hand. By the time it gets to the batter, a 95-miles-per-hour pitch has slowed to 87 miles per hour.

Grape

A grape's terminal velocity is 65 miles per hour—not enough momentum to do any damage even if it hit your head. However, the world record for catching a grape *in your mouth* is 788 feet, set by Paul Tavilla in 1988.

Conclusion: If you see a grape falling, first make sure it's a grape and not something more solid, and then open wide!

Soccer Ball

A soccer ball is relatively large and light, a slow combination for falling objects. It would max out at 54 miles per hour if someone threw one from the top of the Empire State Building. Soccer players regularly kick them faster than that—the record is 132 miles per hour—and go to enormous effort to place their heads in front of them, suffering nothing more than a headache and the loss of a few brain cells as a consequence.

How high would the ball bounce? A soccer ball's coefficient of restitution (COR) (how much energy an object retains after it bounces off a given material, in this case your head) is 0.85. If it hit your head, it would bounce back to the fourth story.

Conclusion: Bouncy, but not lethal. (If you're looking for even bouncier, try dropping a Super Ball. It has a terminal velocity of 70 miles per hour—also not lethal, but with a COR of 0.90 it would bounce 80 feet high if you dropped it from the skyscraper.)

Ballpoint Pen

It depends on the pen. A ballpoint pen without a shirt clip will tumble as it falls and go too slowly to do any damage. If, on the other hand, it's a steel pen with a shirt clip, it would drill that hole in your head that the penny was supposed to. Why?

The shirt clip would act like the feather fletching on an arrow and keep the pen pointed down. Not only would it accelerate to 190 miles per hour, it would hit your head as a rod—and rods are great for puncturing because they carry extra momentum without adding drag (which is why anti-tank ammunition is rod shaped).

Conclusion: Thanks to its "fletching" and its rod-momentum bonus, a falling pen with a shirt clip will puncture your skull and pierce your brain. Result? If falling from the top of a skyscraper, a pen is as mighty as the sword.

Blue Whale

A blue whale holds the world's free-fall speed record for all life forms. Or at least it would if it could find a ride to the top of the atmosphere. Weighing in at 420,000 pounds, a blue whale has the highest falling terminal velocity of any animal that has ever lived. From any elevation greater than 4 miles, a falling whale would break the sound barrier at sea level. From the top of the Empire State Building, a whale would reach 190 miles per hour.*

That spells trouble if you were to try to catch it. You would be flattened. But it's actually worse than that.

If a whale struck the ground it would "splash," meaning its skin would not be able to contain the outward expansion of its guts. The same thing would happen to your body underneath the falling whale. Your skin would fail to contain its contents. So after smashing (and splashing), the whale's insides would mix up with yours.

Conclusion: messy.

---

*Yes, the same as the pen—190 miles per hour is the maximum amount that Earth's gravity can accelerate anything from the height of the Empire State Building.

This Book

If someone should discard this book from the top of the Empire State Building—we know, probably the most unlikely scenario here—it would max out at 25 miles per hour and take more than thirty seconds to complete its fall.

Conclusion: If you have ever angered a strong-armed librarian, you might have been hit by a book traveling faster than 25 miles per hour. Startling, but not deadly.

What Would Happen If...
You *Actually* Shook
Someone's Hand?

ONE OF THE worst things you can do for your health is to shake someone's hand. Hands are our primary disease transmitters, which is why the Centers for Disease Control is a big proponent of the fist pound. Disease alone, though, does not begin to take into account how dangerous your next handshake could be.

That's because you have never *actually* touched another person's hand—even if you're one of those firm shakers—because of something called atomic repulsion. If you *actually* touched the hand of the next person you shook with, the results would be disastrous.

Every atom that makes up your palm (and everything else) has negatively charged electrons that circle their nuclei. These electrons repel one another just like the north poles of your fridge magnets, except that, unlike your fridge magnets, electrons *really* don't like to touch one another.

They repel with such force that you have not actually touched anything in your life. Right now your butt is not actually touching your chair, but hovering above it. Smash a nail with a hammer, and the hammer and nail won't actually touch each other either.

To force two atoms together, you need more pressure than your hand, a hammer, or your butt can provide.

In nature, this kind of pressure is found in the center of stars. Our sun generates heat by pushing hydrogen nuclei together in a process called nuclear fusion.

The only way to create that kind of pressure here on Earth is through explosives.

To truly shake your friend's hand, to truly touch your atoms to his, you would have to bake your hands into a nuclear bomb and set it off. (Note: This is potentially risky. Make sure you have adult supervision.)

Unfortunately for you, your friend, and the city you happen to be located in, the most common molecule in human skin is hydrogen, and when hydrogen nuclei fuse, they release an *enormous* amount of energy.

In getting your two hands to *actually* shake, you have just detonated a medium-size hydrogen bomb.*

Everyone within twenty miles would suffer third-degree

---

*Small technicality: We're cheating a little bit here. The pressure and heat created inside a nuclear bomb do not last long enough to force hydrogen together. To make an H-bomb, physicists use isotopes of hydrogen (deuterium and tritium) made in nuclear fission reactors and then place those isotopes inside a nuclear bomb. The only way to do that to you and your friend's hands would be to process you both in a fission reactor *or* have you shake hands inside a star. However, the logistics of both are complicated, so for the sake of convenience we're skipping that step.

radiation burns and nerve damage. People within six miles would get that plus their houses blown down. People within three miles would get that plus an air blast powerful enough to destroy skyscrapers, and everyone within two miles would get all that plus they would be engulfed in a gigantic fireball.

For you and your friend it would be over quickly. The first thing you would see would also be the last. That's because the flash of the bomb is blinding. Literally. The light would burn out your retinas like an overexposed photo, then vaporize your eyeballs and optic nerves.

Accompanying the flash is a veritable smorgasbord of electromagnetic radiation. To get a sense of its effect on you, imagine this: If you stepped into a microwave, your water molecules would be tickled into moving faster and would heat up. Eventually your liquids would turn to steam and expand. When water expands under pressure, like the pressure of your veins, it explodes. You would coat the inside of the microwave. Except a microwave offers only a sampling of low-powered electromagnetic radiation. An H-bomb gives you the entire photon radiation buffet: infrared, visible, ultraviolet, X-ray, and gamma ray.

The photons would blast your vaporized body and destroy the atomic bonds holding your molecules together—breaking you down into your component atoms.

And then it would get worse.

Your molecules would no longer be attached, but they would still be clumped up like billiard balls. Then along comes the photon cue ball. The photons would hit your atoms and disperse you into an area the size of a high school gymnasium.

Next come the particles. These are your slower-traveling neutrons and electrons, and it's the neutrons that should particularly concern you. They would go after your individual atoms and transmute your nuclei—turning your last remains radioactive.

The slowest traveling effect of the bomb is its supersonic shock wave. That blast of turbulence would push your transmuted, radioactive, ionized-plasma body at high speed, turbulently mixing your atoms with the expanding hot plasma cloud of everything else that used to be you. Eventually you would sprinkle back down to Earth as 10,000,000,000,000, 000,000,000,000,000 separate atoms. Roughly.

What Would Happen If . . .
You Were the Ant Under the Magnifying Glass?

Any kid knows that a magnifying glass can scorch an ant. Fortunately CVS doesn't sell magnifying glasses big enough to heat blast a person, but with enough people and a lot of mirrors you could give someone a lot more than a sunburn.

In Arthur C. Clarke's "A Slight Case of Sunstroke," a president conceives a diabolical plan to counter a crooked referee. He gives 50,000 soldiers a free ticket to a soccer game and a two-foot reflective program. The soldiers think they're being given a novel way to boo, but the president has deadlier intentions, and after a particularly egregious call they all use their programs to reflect sun onto the ref. The power of 50,000 reflecting mirrors burns the referee alive.

The story is fictional but the theory is surprisingly sound, and executed properly it would take a lot less than 50,000 fans.

Arthur C. Clarke is not the first person to think of using sunlight as a weapon.

According to legend, Archimedes burned enemy ships by having 129 soldiers reflect sun onto them with their brass shields. With the technology available to Archimedes it's clear this didn't happen, though a study by MIT showed it's theoretically possible.

Although focused sunlight has never killed a human, it does kill thousands of birds every year. Solar farms in the Mojave Desert use garage-door-size mirrors to focus sunlight into a 1,000-degree bird-frying beam.

The biggest issue with weaponizing sunlight, which the solar farms solve with movable mirrors and computer algorithms, is focus.

Once you have more than ten or twelve squares of light on an object it becomes very difficult for each person to aim their light. Nobody knows which square they're aiming.

The U.S. Air Force has solved this problem. Included in their survival kits is something called a signal mirror, and it's a powerful tool for a downed pilot.

A small mirror reflecting sunlight can flash distress signals that can be seen for miles. The trick is aiming the mirror's light, and air force signal mirrors use reflective beads to place a red dot, like a sniper's scope, where your reflection is pointing.

The mirrors are surprisingly effective. In 1987, a father and son rafting down the Colorado River in the Grand Canyon had an accident and used a signal mirror to successfully signal SOS at an airliner passing 35,000 feet overhead. Attaching signal mirrors to each fan's mirror solves the focus issue, and once you do that, the heat scales quickly. A standard one-foot-by-one-foot bathroom mirror receives one hundred watts of energy from the sun, and each mirror reflects exactly as much heat as it receives. So one mirror reflects the heat of that mirror, two mirrors the heat of two mirrors, and so on.

So if you were the referee in the Arthur C. Clarke story and you made a couple of bad calls, here's what we think you could expect. If the fans brought basic mirrors, as they do in the book, you would not have much to worry about. The light would be too diffuse to do anything other than make you a bit warmer, and you would have plenty of time to scamper off the field.

If, on the other hand, it was a sunny day and 1,000 fans attached signal mirrors on top of their bathroom mirrors, you would need to be very concerned, because they could collectively direct 100,000 watts onto your chest. That's enough heat to boil a 200-pound person in a few minutes, but you would die long before you began boiling.

A well-stoked fire is 800 degrees—if you stuck your hand close to one you would have to pull it away, and in the glare of 1,000 mirrors the temperature would be even hotter, in the neighborhood of 1,000 degrees.

Your cells work in only a narrow band of temperatures. They're happy at 98.6 degrees, and even a 2-degree rise will make you uncomfortable. A 10-degree rise in your temperature is lethal.

Fortunately, you have developed a number of ways to keep your inner temperature cool in scorching heat. Sweating, dilating your veins, and your body's insulation can keep you alive in rooms over 200 degrees for a few minutes.

But in extreme conditions, everything happens too fast for your defense mechanisms to do any good.

If 1,000 perfectly aimed mirrors reflected their beams on you, you would be dead before you took two steps. You would not combust right away because you have so much water in you that you are like a soaked log, but as soon as you took a breath, the tender skin in your throat would burn and scar, disabling it permanently. You would suffocate if you were able to last another minute or two, but, not to worry, there's no chance of that.

Instead your inner body temperature would rocket up the required 10 degrees, your brain cells would stop working, and your proteins would denature (what physicists call cooking).

Nothing in your body works without your proteins ferrying energy, so you would be dead meat.

But your body would keep cooking until it was fully dehydrated and then you would burst into flame. The fire would gradually move through you until there was nothing left but bones and teeth.

Crematoriums heat up to 1,500 degrees and take two and a half hours to fully transform someone to ash, so unless the fans were truly dedicated, at least a few of your teeth and scorched bones would remain on the field.

Perhaps, as in "A Slight Case of Sunstroke," your death would be followed by a short moment of silence, a "new and understandably docile referee," and a comeback victory by the home team.

What Would Happen If . . .
You Stuck Your Hand
in a Particle Accelerator?

In July 1978 a Russian scientist named Anatoli Bugorski was inspecting Russia's most powerful particle accelerator (a machine that speeds subatomic particles to near the speed of light), called the U-70, when the main particle beam hit him in the back of the head and passed through his nose. He felt no pain but reported seeing a flash "as bright as a thousand suns." Russian doctors rushed him to the hospital expecting him to die from radiation poisoning, but, other than some facial paralysis, the occasional seizure, a touch of radiation sickness, and a small hole through his head, he was fine and went on to finish his PhD.

Does this mean you could stick your hand in Europe's new Large Hadron Collider (LHC)? Would you get a cool scar but otherwise be unharmed? No. Unfortunately for you and your hand, the Russian U-70 accelerator had less than 1 percent of the power that the LHC does.

The LHC is the most powerful particle collider in the world. It accelerates protons around a 17-mile loop to 0.99999999 c (7 miles per hour less than the speed of light) and smashes them together in the world's greatest demolition derby. It's so powerful that a small but vocal community expressed concern that the smashing particles would create a black hole large enough to consume Earth (see p. 197 for what would happen if it did).

The beam is composed of 100 billion protons, which, when accelerated to near light speed, carry a huge amount of energy—similar to a 400-ton train traveling at 100 miles per hour.

The beam carries so much energy it can drill a hole through 100 feet of copper in a millisecond—which is why most accelerators are pointed into the ground, ensuring that if a malfunction occurs a killer beam is not shot through a city.

So you can see right away that there will be a few issues with sticking your hand in the beam, but let's say you ignored the warning signs and did it anyway. The first problem? Your ears.

Carbon-fiber jaws guide the beam's path. If the beam wanders, it strikes the carbon fiber, and for you the sound would be as loud as if you were standing in front of concert speakers. Then, when scientists are done experimenting, the beam's energy is dumped into a graphite block used as a proton trap, which would sound like a 200-pound TNT explosion—loud enough to blow out your eardrums.

In other words, wear some earplugs. But, really, blown-out eardrums would be the least of your concerns. A bigger problem would be the power of the beam.

The protons would pass through your hand as if nothing were there at all. The beam is small, about the width of the lead in a pencil, and traveling so fast it would punch through your hand painlessly. There's a good chance it would miss your bones and leave your hand fully functional, but that is only if you kept your hand very, *very* still.

The U-70 Russian reactor was not only less powerful than the LHC, it was also only a single shot, so Bugorski had only one hole in his head. The LHC is more like a proton machine gun—in two seconds it fires nearly three thousand shots. If you pulled your hand away after the first pulse, the beam would cut your hand in half.

So don't do that.

As the beam passed through your (hopefully) steady hand, another far more troubling process would take place. Particles traveling as fast as these, by their nature, are accompanied by intense radiation. Even if you were many yards away from the beam you would be dosed with the equivalent of a full chest X-ray.

Exactly how much radiation you would receive if the beam *hit* you, though, is actually difficult to say. The beam itself carries a gargantuan amount of radiation, enough to kill you quickly (and many times over), but the vast majority of the radiation would miss you because although you might think your hand is solid, at an atomic level there's actually quite a bit of space.

If an atom in your hand were enlarged to the size of a football stadium, then a marble sitting on the fifty-yard line would be the nucleus. Because the radiation bullets fired at you are also quite small, most of them would miss, saving you from instant death. Unfortunately, only most of them would miss. You would probably be hit with just enough radiation to kill you slowly and painfully.

In the end, because Bugorski nearly died of radiation poisoning despite the accelerator being less than 1 percent as powerful, we can be confident that a beam from the LHC would kill you. The particles created when the beam struck your hand would irradiate and poison your entire body with at least 10 sieverts of radiation, and your experience would likely mimic what two workers at the Tokaimura nuclear processing plant went through after an accident in 1999.

Hisashi Ouchi and Masato Shinohara were creating a small batch of nuclear fuel when they miscalculated the recipe and their mixture went critical. Even in lethal radiation exposures victims don't always feel terrible right away. The symptoms can take a few hours to set in. But in extreme exposures—like yours, Ouchi's, and Shinohara's—there's no delay.

Just after the beam pierced your hand, your vision would turn blue, the result of radiation passing through the liquid of your eyeball faster than the speed of light. The speed of light is 30 percent slower in water than in a vacuum and it produces an electromagnetic shock wave, called Cherenkov radiation, that appears blue. Both Ouchi and Shinohara reported that the room turned blue despite security cameras showing no change in color.

Besides appearing to change its hue, the room would feel hot even though the actual temperature would remain unchanged as the beam's energy heated you.

You would also feel nauseated almost immediately as the radiation attacked the lining of your stomach. Your skin would be severely burned, you would have trouble breathing, and you might lose consciousness.

Your white blood cell count would drop to near zero, preventing your immune system from functioning, and the damage to your internal organs would slowly progress. Doctors would be able to treat your symptoms but would not be able to do anything about your irradiated organs. Depending on the exact dosage you received and the progression of the damage, you would die within four to eight weeks.

The hole in your hand, however, would be small enough that it should heal in time with only a small scar.

What Would Happen If...
You Were Holding This Book and It Instantly Collapsed into a Black Hole?

When the Large Hadron Collider was first proposed a vocal few were concerned that the smashing atoms would create a small black hole that would consume Earth. Fortunately, that didn't happen. Creating a black hole is beyond our capabilities, as things stand. And that's a good thing, because even small black holes should be avoided. If this book collapsed into a black hole, a few things would happen—all of them bad.

Anything can become a black hole if it's squeezed small enough. Most things don't, though, because there's nothing to squeeze them down to size. The only thing we know of that can crush an object small enough to create a black hole is a massive star's own gravity.

Everything has its own gravitational field, but it takes a truly enormous star—at least twenty times the size of our sun—to have gravity strong enough to crush it down small enough to create a black hole.*

It's *possible* that during the big bang such huge compressive forces were created that objects smaller than massive stars—objects the size of this book, say—were crushed into black holes.

That's a long way of saying that while it's unlikely that this book will turn into a black hole when you're done reading it, it isn't impossible.

You're going to want to back away if it does.

Assume this book has a mass of about a pound. Collapsed into a black hole it will still have that same mass, only it's going to be very, *very* small. About 1 *trillion* times smaller than a proton—which itself is just a small part of an atom.

Stephen Hawking calculated that black holes are not perfectly black. Instead, they leak out Hawking radiation until they die. For big black holes that takes a long time (the black hole at the center of the Milky Way will take 1 googol years to evaporate), but this tiny *And Then You're Dead* black hole is going to vanish a split second after it's created.

It wouldn't go quietly. In that split second the book would explode with five hundred times the energy of the Hiroshima bomb. It would emit a bright flash and bombard the area with a full spectrum of light including X-rays and gamma rays, and the air would become ionized, heat up, and glow. A massive

---

*How small is the singularity and what does it look like? Because of the nature of black holes, where no light can escape, there is no way for physicists to confirm any theories regarding the inside of black holes. So we don't know.

shock wave, powerful enough to knock down buildings, would spread for miles.

You and your surrounding area would be totally destroyed, but, fortunately, the information in this book would not be.

According to the latest theories published by Stephen Hawking and others, the information inside black holes *is not* totally destroyed; it's just in a language that we have no idea how to read.

Unfortunately, it could be many thousands of years before physicists are able to read the data leaked from black holes, and by that time English and all other current languages are likely to have long been lost.

So, though it's unlikely that this book will ever be converted into a black hole and therefore improbable that you would be blown apart, irradiated, vaporized, transmuted, and ionized, it's not impossible, and thus we feel compelled to speak to the future physicists who will sift through the ancient wreckage in the only way we feel confident they will be able to understand.

To these future beings, we say: ☺

What Would Happen If . . .
You Stuck a Really, *Really* Powerful Magnet to Your Forehead?

Take a kitchen magnet and put it up against your forehead. What happens? Nothing, right? Not even a tingle.

That's because you're impervious to the pull of kitchen magnets, and in fact you're impervious to the pull of the strongest magnets we have on Earth. The most powerful magnet created by researchers registers at 45 tesla—a kitchen magnet is 0.001 tesla—and though it would be strong enough to levitate you (we'll get to that), it would be harmless.

But what if you decided to look elsewhere for a magnet? The biggest kitchen magnets in the galaxy are rare versions of neutron stars called magnetars, which register at an atom-deforming 100 *billion* tesla.

Neutron stars are stars that have gone supernova but were not big enough to create black holes, so instead were crushed by their own gravity into superdense gigantic nuclei. A magnetar's initial extremely fast rotation gives it an enormously powerful magnetic field.

Magnetars are such fantastically strong magnets that if the moon were replaced by a magnetar, it would wipe out every credit card on Earth. This powerful magnetism also makes it one of the most destructive stars in the galaxy. If we wrote this book forty years ago we would have had no idea magnetars existed and thus would have sworn that you couldn't die from magnetism in our galaxy. Then in 1979, a magnetar had a starquake, hit us with 100 times more gamma rays than our satellites had ever measured, and alerted us to their presence.

In 2004, an even more powerful one struck. A magnetar 50,000 light-years away released as much energy as our sun does in 250,000 years. The gamma ray burst fried satellites and altered Earth's magnetic field. If you were unlucky enough to be within a light-year of a magnetar when it had a quake, you would be X-rayed to death.

If the magnetar weren't experiencing a starquake you could get closer, but once you got within 600 miles the extreme magnetism would become a problem.

You probably don't think of yourself as a magnet, but in truth you are—you're just a pathetically weak one. Water—which makes up 80 percent of our bodies—is a diamagnetic material, meaning it's repelled by both the north and south poles of a magnet. That means a strong enough magnet would repel you with enough force that you would float.

Scientists once floated frogs within 10-tesla magnetic fields—magnets five times stronger than an MRI machine (no frogs were harmed in this experiment). Scientists could float you if they made a 10-tesla magnet large enough for you to fit inside.*

Unfortunately, you wouldn't float harmlessly above a magnetar because a number of your body's processes are affected when you're exposed to magnetism 100 billion times stronger than an MRI machine.

Right now your atoms look like beach balls as your electrons circle their nuclei. That's good; it's how they should be. But your electrons are also magnetic. Once you were within 600 miles of a magnetar, the magnetic pull would become strong enough to tug at your electrons and give them an elliptical orbit. So instead of your atoms looking like beach balls, they would look like cigars. That's not good.

Without spherically shaped atoms, your proteins unfold and the bonds binding your atoms into molecules break, instantly splitting you into billions of independent atoms. Instead of H_2O, for example, you would now have two Hs and an O.

This would be profoundly fatal.

If someone were looking at you from a passing spaceship when the magnetism first pulled your molecules apart, you would appear as a shimmering human-shaped gas, but that would not be your final form.

Different atoms of yours have different magnetic properties, so some parts of you would be pulled toward the magnetar

---

*We are convinced it would be totally harmless and volunteer to go first.

faster than others, stretching your "body" out. Then the magnetar's gravity would begin tugging at you.

Magnetars are small—about the size of Manhattan—but incredibly dense, giving them an enormously strong gravitational field that would pull you toward the star, overpowering the star's magnetic repulsion. In your stretched-out form, you would accelerate toward the magnetar.

To any nearby observers, your last remains would appear as a long wisp of gas like smoke from a chimney, accelerating toward the magnetar until you augered into it, whereupon your atoms would be transmuted into a smear of neutrons and crushed down to the size of a single red blood cell.

What Would Happen If . . .
You Were Swallowed by a Whale?

THE OLD TESTAMENT tells the story of Jonah, a disobedient prophet who is swallowed by a whale and spends three days in its belly before being spat out on a beach, healthy and whole and in the presence of a few incredulous witnesses. Amazed by the feat, onlookers lead the reprobate city of Nineveh to repent their sins in the face of such a miracle.

Jonah may indeed have had some outside assistance because according to marine biologists, entering the belly of a whale is a risky endeavor. Your best bet to get a ride inside the stomach of one of these beasts is to find a sperm whale. Most whales eat microscopic organisms like plankton, and so their throats are only four or five inches wide. If you happen to find yourself inside the mouth of a blue whale you would be too big for it to swallow and your journey would likely end with a crushing blow from its 6,000-pound tongue.

Sperm whales, on the other hand, eat larger prey like giant

squid and have been known to swallow the 400-pound animals whole. So it could, in theory, swallow you. But even if you escaped the whale's teeth and tongue, you would find yourself in the first of the whale's four stomachs and facing another set of problems.

Whales are flatulent creatures. The only gas you would find in its belly is methane, not oxygen. And while it's not toxic, it is a natural asphyxiate. Most untrained humans can't hold their breath much longer than thirty seconds. A lack of oxygen is fine for most of your tissue, which can go for hours without a resupply, but your brain is another matter entirely. Once the leftover oxygen in your bloodstream was exhausted, brain cells would begin to die immediately. Unless a higher power stepped in, irreversible brain damage would begin within four minutes and complete brain death would follow a couple of minutes later.

You would also have the whale's stomach muscles to contend with. Because sperm whales don't chew their food, they rely on the muscles in their first stomach to squeeze prey down to size. So before you had a chance to be dissolved in powerful stomach acid, the muscles of its stomach would squeeze you into something resembling chunky peanut butter.

There is some good news, though.

Sperm whales have the most expensive poop in the world. Their bile duct secretion, called ambergris, is a prized commodity in the perfume industry. A one-pound chunk is worth around sixty thousand dollars.*

---

*The next time you're at the beach, keep an eye out for a hard, smelly, cream-yellow or dark brown "rock" that looks like a giant piece of earwax. It could make you rich.

So it's possible that after being suffocated, crushed, dissolved, passed through a thousand feet of intestines, and pooped out the back end of a whale, your remains will wash up on a beach somewhere, where a lucky sunbather will pick up your waxy, manure-smelling, ambergris-covered corpse and sell it for a small fortune. And from there, things really start looking up. After the perfumer is done with you, not only will your corpse smell significantly better, your final resting place won't be a hole in the ground or the bottom of the ocean, but on the back of a lady's neck as a gentle spray to improve her scent.

Considering the alternatives, you too may believe in divine intervention.

What Would Happen If . . .
You Took a Swim Outside a Deep-Sea Submarine?

In January 1960, two navy divers piloted a specially de-signed submarine to the deepest part of the ocean, the Mari-ana Trench, a crevasse in the sea floor off the island of Guam roughly 7 miles deep. It took Don Walsh and Jacques Piccard nearly 5 hours to reach the bottom. They had just 20 minutes to survey the landscape before a crack in their window prompted a hasty retreat to the surface.* During their short visit they conducted some scientific experiments and made a few observations. They did not, however, exit the sub for a swim.

But what if they had?

---

*What if their window had cracked all the way through? The force of the ocean pushing the water through their window would have created a stream powerful enough to cut through both of them *and* the other side of their sub. And then they would have been crushed.

Anyone who has swum to the bottom of a pool can tell you about the squeeze felt under a few feet of water, especially in the ears. But at 7 miles that pressure is multiplied roughly a thousandfold. Sitting at the bottom of the pool your body is under 12 feet of water, equivalent to 5 pounds per square inch of pressure. At the bottom of the Mariana Trench you would experience 15,750 pounds per square inch. Surprisingly, this crushing weight would not smash your body, or at least not your *entire* body. That's because, with a few exceptions, you're mostly just water, and water is incompressible. Unfortunately for you, you're not *all* water—and it's those gases that are going to be a problem.

The instant you stepped out of the sub your eardrums would blow out, your nasal cavities would collapse, and your throat would cave in. None of that's good, but the real problem would be your chest, which would collapse as your lungs were crushed to the size of Ping-Pong balls and refilled with water. Every air pocket in your body would crunch down until you were a tightly packed humanoid-shaped flesh chunk.*

It's probably just as well, considering your appearance, but you would likely never be seen again. Your corpse would not float to the surface because every air pocket inside your body would have collapsed, and your body would decompose slowly because bacteria don't do well in freezing temperatures. Most

---

*It would also be quite chilly. Even though on the surface above the Mariana Trench you would be comfortable in a bathing suit, cold water is denser than warm water, so it sinks. Stepping outside the sub you would face 34-degree water (which would kill you in around 45 minutes), but because your face would also be smashed in we don't think you would care about the temperature.

likely your flesh would be consumed by the various creatures living at the bottom of the ocean, and your bones would be eaten by the gloriously named bone-eating snot flower, which normally lives on whale bone but would likely make an exception in your case.

What Would Happen If...
You Stood on
the Surface of the Sun?

T**HOUGH HUMAN LIVES** are fragile, the matter that makes up our bodies isn't. Whether you jumped into a volcano or were hit by an asteroid, at least a few of your atoms would remain. However, if you're dead set on destroying every last piece of you, down to the atomic level, look to the sun.

The fastest, though not the most fuel-efficient, way to the sun is simple: Allow yourself to fall into it.* Right now Earth is orbiting the sun at 67,000 miles per hour—and orbiting is nothing more than falling into an object while traveling sideways so quickly you miss it. So to reach the sun, all you need to do is stop your horizontal velocity.

First, you would need to exit Earth's gravity—around 1 million miles away (four times farther away than the moon) should

---

*It's not fuel efficient but it's still quite green because you're removing those fossil fuels from Earth entirely. So your spaceship is better for the environment than any hybrid.

be enough—and then fire retro-rockets that slowed your orbit around the sun from 67,000 miles per hour to zero.

Then you would start accelerating. By the time you reached the sun you would be traveling at 384 miles per second—or 1.4 million miles per hour, by far the fastest any human has ever gone—and you would reach it in just 65 days. The first 64 days of your trip would pass smoothly. You would need an X-ray and a heat shield—we recommend the carbon fiber shield used by the (unmanned) *Solar Probe Plus*, which NASA will be sending to the sun. That shield is good enough that even when the temperature hits 2,500 degrees (within four hours of travel time from the visible surface of the sun), the inside of your ship would be at room temperature.

Unfortunately, over the last four hours of your descent, the temperature would climb well past the shield's capability.

The sun's magnetic field heats the outer atmosphere, called the corona, to 2 million degrees.

Because you would still be in a vacuum, you would feel only the 10,000 degrees of radiating heat of the surface at first—but that's still enough to vaporize your heat shield, your spacecraft, and you.

Then, after spending some time in the sun's corona, what was left of you would slowly be cooked to 2 million degrees and you would be turned into a fourth state of matter—a highly ionized plasma. In that condition the solar magnetic field would grab and stretch you into a thin spaghetti string around the sun, then bend and twist you into arcs of glowing light—a beautiful sight for any space telescope focused on the sun.

You may also return home. Once you were shredded down to size, the solar magnetic fields would fling you into space fast enough to cover the 100-million-mile distance to Earth in a couple of days.

So far what we have described is all feasible. You could be turned into a highly ionized plasma if NASA made it a priority. But let's depart from reality for a second and make an improvement to your heat shield that would allow you to travel past the corona and make it to the visible surface of the sun.*

Once you got past the corona and to the visible surface, the temperature would actually cool off to a relatively mild 10,000 degrees as you left the near vacuum of the corona and made it into the sun's atmosphere.

Assuming your heat shield was still holding strong, the first thing you might notice would be the sound. In space no one can hear you scream, nor can anyone hear the deafening roar of the sun. If sound could travel perfectly through space, the sun would sound like a revving motorcycle to us on Earth. On the surface, though, when the sun's gas bubbles popped, the noise would be deafening—one hundred times louder than standing in front of concert speakers. That's enough power to blast out shock waves that destroy the alveoli in your lungs.

Let's assume you came prepared with a fantastic insulator, though, and you made it to the center of the gas star.

The greatest difference between our sun and Jupiter is not the ingredients each is made out of, but *how much* of each

---

*The sun doesn't have a surface; it's all gas, like Jupiter, but it has a layer of ionized gas too thick to see through—so that's what we are referring to as the sun's surface.

ingredient (helium and hydrogen, mostly) there is. The sun is a thousand times as massive as Jupiter, which means the temperature and pressure in the center are so great, nuclear reactions begin to take place.

Nuclear reactions are dangerous to stand near, or in this case, participate in.

Inside the sun, the temperature reaches 27 million degrees with a pressure of 250 billion times the surface of the Earth. That's bad for the parts of you made of hydrogen—meaning most of you. In this heat those hydrogen atoms would move so fast they would ram into one another, eventually fusing together to make deuterium and tritium isotopes of hydrogen. The isotopes would then be slammed together to make a helium nucleus. The end result: You would now be a slow-motion hydrogen bomb.

However, it should be noted that as well as the sun produces heat, you do an even better job. As you sit on the couch converting food into energy you're producing more heat by weight than the sun does. What makes the sun so hot is its enormous size. If you were as massive as the sun, the chemical energy you produced would make you the hottest star in the galaxy.

So if you somehow did make it inside the star, for the briefest moment—just before you were irradiated and vaporized—you would make the sun just a little bit warmer.

What Would Happen If . . .
You Ate as Many Cookies as Cookie Monster?

WHEN IT'S EMPTY, your stomach is about the same size as your fist and frustratingly limited during feasts. Fortunately, its walls are stretchy, so when cookies are offered for dessert you're able to eat one, two, three, four . . .

But a stomach can't stretch forever, and the muscles involved in food swallowing are powerful enough to force more cookies into your stomach than it can handle.

That can lead to an issue.

The expert on cookie stuffing is, of course, Cookie Monster. For the record, he's appeared in 4,378 *Sesame Street* episodes, and an unscientific survey reveals that he eats approximately 3 cookies per episode. That's a total of 13,134 cookies consumed. Although that's a big number, when you spread it out over a 45-year TV run it's perfectly safe.

But what if you were to give Cookie Monster a run for his money and eat all those cookies in one sitting?

Satiety is the medical term for being stuffed. It's a complicated process that involves not just the quantity of food consumed but also the kind of calories it contains. Different calories trigger different responses—protein and fiber increase satiety while carbohydrates and fats have less effect.

The signal from your stomach to your brain is also a bit delayed—it can take fifteen to twenty minutes for your brain to get the message, which means the faster you eat, the more cookies you can jam into your stomach before you realize there's an overfull issue.

Most people are typically satiated after about 25 cookies' worth of food—more or less what Cookie Monster eats in a single frenzy. Of course, stomachs are stretchy, so 25 is not the physical limit, and competitive eaters have a few tricks to stretch it.

First, a thin physique helps. Exactly how you can stay thin while eating 11 pounds of cookies is a bit of a paradox, but it's true that with less fat in the way your stomach has more room to expand outward.

Second, in preparation for eating all those cookies, you could to do some limbering up. Eating low-calorie, highvolume food (like grapes) the night before would stretch your stomach and help make it ready to stretch again.

For Cookie Monster and you, the 60-cookie mark is where things would start to get difficult. (To be clear, we're talking about an average-size chocolate chip cookie, not one of those cookie behemoths.)

Unless you have eaten 60 cookies many times (and thus suppressed your gag reflex), your stomach will revolt and you will vomit. But that's a good thing: 60 cookies equals roughly 4 liters of food, and that's approaching your stomach's breaking point.*

We know the physical limit of the stomach thanks to the German physician Algot Key-Aberg, who in the late 1800s attempted to cleanse a patient of an opium overdose by pumping water into his stomach. Unfortunately, the patient's drug use suppressed the normal vomiting response and his stomach broke like an overfilled water balloon, killing him on the operating table.

This event piqued Key-Aberg's curiosity and he began experimenting to determine the true capacity of a stretched human stomach using corpses. He concluded that the typical stomach can hold 4 liters of food before eruption. (Imagine two party-size sodas next to each other. If you eat or drink more than that, you are approaching what we will call here the stomach eruption limit.)

This limit applies to all of us, except, however, for a gifted few. A small number of people have publicly passed the 4-liter mark. Depending on your training, or whether you received the genetic gift of a flexible stomach, it *is* possible to eat more. Joey Chestnut, the reigning hot-dog-eating champion, once ate 69 hot dogs in 10 minutes. That's approximately 9.5 liters of food—or 130 chocolate chip cookies.

But let's say you lack any genetic stomach gifts. You would

---

*People with binge-purge disorders are at particular risk for this injury because their bodies have become accustomed to overfilled stomachs and their gag reflex has been suppressed. A fashion model in London once ate nineteen pounds of food in one sitting—the equivalent of eighty cookies—and died from a stomach rupture.

begin to run into real trouble at the 90-cookie mark, or about 6 liters of food.

The weakest part of the stomach is the lesser curvature. If you imagine your stomach looking like a kidney bean, the lesser curvature is the area that bends inward. This is where the cookies would initially break through.

The body's innards have little defense against the bacteria that live on cookies. Clostridium perfringens,* otherwise known as gas gangrene, starts to grow in your gut as soon as the cookies bust out. It destroys live tissue and produces gas that explodes and distributes dead and rotting material throughout your guts.

In response to the massive bacterial invasion, your immune system would send an overwhelming amount of chemicals to the infected area. This is known as septic shock and constitutes a body's defense to widespread infection. It can be so overwhelming that the response itself can kill you. How? Inflammation, blood clots, and decreased blood flow. The pulse picks up to try to get more blood to vital organs, the body temperature drops, often dangerously, and gas gangrene can appear.

This infection lives within a protective cocoon of dead tissue, outside the reach of white blood cells and antibacterials. Once the body progresses to this stage, which it would do rapidly, you would be unlikely to survive even with skilled medical care. Within an hour your heart would not receive enough oxygen to keep beating and you would go into cardiac arrest followed quickly by total brain death.

---

*Do not search Google Images for this term.

All that being said, it's possible that you would actually die *before* that. Remember that a typical unstretched stomach is the size of your fist. After you have filled it with 6 liters' worth of cookies, that stomach would now be more than 20 times its normal size. This begins to get in the way of other bodily functions. The vein that runs below the stomach and returns blood from the bowels to the heart would be pinched shut.

Then there's the breathing problem. The stomach's upward growth can also compromise your lungs. At 20 times normal size, your stomach would have grown into your lung space and you would be suffocating on cookies.

Between suffocation, stomach explosion, and bowel death from oxygen starvation (never mind septic shock), the medical battle to save you would be very close, and in the end would probably depend on the amount of gas produced during digestion. After 60-some-odd cookies, the gaseous side effects of digestion might push the pressure of your stomach beyond its physical capacity. It could explode violently and distribute its fatal chocolate chip cookie content throughout your innards.

In other words, death by burping.

References
and Further Reading

Nᴏᴛ ᴇɴᴏᴜɢʜ ɢʀᴜᴇsᴏᴍᴇɴᴇss for you? We've compiled some of our favorite sources below. Here you can find plenty more details on factory mishaps, shark attacks, and air force experiments if the previous pages didn't satisfy.

You Were in an Airplane and Your Window Popped Out?

Average human body proportions

http://www.fas.harvard.edu/~loebinfo/loebinfo/Proportions/humanfigure.html

The actual story of the British Airways pilot who was sucked out of his windscreen

http://www.theatlantic.com/technology/archive/2011/04/what-to-do-when-your-pilot-gets-sucked-out-the-plane-window/236860/

You Were Attacked by a Great White Shark?

Unprovoked fatal shark attacks in the United States

https://en.wikipedia.org/wiki/List_of_fatal,_unprovoked
_shark_attacks_in_the_United_States

Treatment for vascular trauma

http://www.trauma.org/archive/vascular/PVTmanage.html

You Slipped on a Banana Peel?

Frictional coefficient of a banana peel

Frictional Coefficient under Banana Skin, https://www.jstage
.jst.go.jp/article/trol/7/3/7_147/_article

Durability of the human skull and the physics of skull fracture

Gary M. Bakken, H. Harvey Cohen, and Jon R. Abele, *Slips,
Trips, Missteps and Their Consequences*, 119

You Were Buried Alive?

Mechanism of death in avalanche victims

H. Stalsberg, C. Albretsen, M. Gilbert, et al., *Vichows Archiv
A Pathol Anat* 414, no. 5 (September 1989): 415
http://link.springer.com/article/10.1007%2FBF00718625

Survival-in-a-closed-space equation

http://www-das.uwyo.edu/~geerts/cwx/notes/chap01/ox
_exer.html

Death by CO_2 buildup

http://www.blm.gov/style/medialib/blm/wy/information
/NEPA/cfodocs/howell.Par.2800.File.dat/25apxC.pdf

You Were Attacked by a Swarm of Bees?

The complete Schmidt sting pain index

Justin O. Schmidt, *The Sting of the Wild*

National Geographic on Smith

http://phenomena.nationalgeographic.com/2014/04/03/the
-worst-places-to-get-stung-by-a-bee-nostril-lip-penis/

Smith's study on honeybee sting pain by body location

https://doi.org/10.7717/peerj.338

You Were Hit by a Meteorite?

Price of a meteorite

http://geology.com/meteorites/value-of-meteorites.shtml

Meteorite impact tsunamis

https://www.sfsite.com/fsf/2003/pmpd0310.htm

You Lost Your Head?

The story of Phineas Gage

Malcolm Macmillan, *An Odd Kind of Fame: Stories of Phineas Gage*

Hydrocephalus case studies

Dr. John Lorber, "Is Your Brain Really Necessary?"
http://www.rifters.com/real/articles/Science_No-Brain.pdf

You Put on the World's Loudest Headphones?

Loudest sounds in recorded history

http://nautil.us/blog/the-sound-so-loud-that-it-circled-the
-earth-four-times

How long would you have to yell to heat up a cup of coffee?

http://www.physicscentral.com/explore/poster-coffee.cfm

You Stowed Away on the Next Moon Mission?

Human exposure to a vacuum

http://www.geoffreylandis.com/vacuum.html

A human exposed to low pressure

https://www.sfsite.com/fsf/2001/pmpd0110.htm

You Were Strapped into Dr. Frankenstein's Machine?

Electric current effects in the human body

http://www.ncbi.nlm.nih.gov/pmc/articles/PMC2763825/

Your Elevator Cable Broke?

Story of Nicholas White stuck in the elevator

http://www.newyorker.com/magazine/2008/04/21/up
-and-then-down

You Barreled over Niagara Falls?

Daredevils of the falls

http://www.niagarafallslive.com/daredevils_of_niagara
_falls.htm

NASA study on fatal fall heights

http://ntrs.nasa.gov/archive/nasa/casi.ntrs.nasa.gov
/19930020462.pdf

The splat calculator

http://www.angio.net/personal/climb/speed

You Couldn't Fall Asleep?

Sleep deprivation on rats

http://www.ncbi.nlm.nih.gov/pubmed/2928622

Randy Gardner story and other neurological findings after sleep deprivation

http://archneur.jamanetwork.com/article.aspx?articleid
=565718

You Were Struck by Lightning?

The book on lightning

Martin A. Uman, *All About Lightning*

Exact timing of lightning strikes and your heartbeat

Craig B. Smith, *Lightning: Fire from the Sky,* 44

Ben Franklin and electrostatics

https://www.sfsite.com/fsf/2006/pmpd0610.htm

You Took a Bath in the World's Coldest Tub?

CERN accident report

https://cds.cern.ch/record/1235168/files/CERN-ATS
-2010-006.pdf

Volume of liquid helium

http://www.airproducts.com/products/Gases/gas-facts/con
version-formulas/weight-and-volume-equivalents/helium.aspx

Supercold temperatures

https://www.sfsite.com/fsf/2010/pmpd1007.htm

You Skydived from Outer Space?

Calculate orbital speed

http://hyperphysics.phy-astr.gsu.edu/hbase/orbv3.html

Falling bodies in the atmosphere

http://www.pdas.com/falling.html

You Time Traveled?

History of the sun

http://www.space.com/22471-red-giant-stars.html

Timeline of the far future

http://www.bbc.com/future/story/20140105-timeline-of-the
-far-future

History of oxygen in our atmosphere

https://en.wikipedia.org/wiki/Atmosphere_of_Earth#/me
dia/File:Sauerstoffgehalt-1000mj2.png

What would it be like to live in the dinosaur era?

http://www.robotbutt.com/2015/06/12/an-interview-with
-thomas-r-holtz-dinosaur-rock-star/

Universal edibility test

http://www.wilderness-survival.net/plants-1.php#fig9_5

Fossils record the past

https://www.sfsite.com/fsf/2015/pmpd1507.htm

You Were Caught in a Human Stampede?

Standing crowd density

http://www.gkstill.com/Support/crowd-density/CrowdDensity-1.html

Past crushes and prevention strategies

http://www.newyorker.com/magazine/2011/02/07/crush-point

You Jumped into a Black Hole?

Falling into a black hole

https://www.sfsite.com/fsf/2015/pmpd1501.htm

Spaghettification

Neil deGrasse Tyson, *Death by Black Hole: And Other Cosmic Quandaries*

You Were on the *Titanic* and Didn't Make It into a Lifeboat?

What happens during a brain freeze

http://www.fasebj.org/content/26/1_Supplement/685.4.short

You Were Killed by This Book?

Bomb calorimeter

http://www.thenakedscientists.com/forum/index.php?topic=14079.0

You Died from "Old Age"?

Gain or lose microlives

http://www.scientificamerican.com/article/how-to-gain-or
-lose-30-minutes-of-life-everyday/

A simple derivation of the Gompertz law for human mortality

http://www.ncbi.nlm.nih.gov/pubmed/18202874

You Were Stuck in . . . ?

Standard atmosphere by altitude

http://www.engineeringtoolbox.com/standard-atmosphere
-d_604.html?v=8.3&units=psi#

Military survival guide, chapter 7—don't drink urine

http://www.globalsecurity.org/military/library/policy/army
/fm/21-76-1/fm_21-76-1survival.pdf

You Were Raised by Buzzards?

Composition of raw meat

http://time.com/3731226/you-asked-why-cant-i-eat-raw
-meat/

The microbiome of New World vultures

http://www.nature.com/ncomms/2014/141125/ncomms6498
/full/ncomms6498.html

Why it's a bad idea to scare a vulture

http://animals.howstuffworks.com/birds/vulture-vomit.htm

You Were Sacrificed into a Volcano?

Geologist actually falls into lava (and survives)

http://articles.latimes.com/1985-06-14/news/mn-2540_1
_kilauea-volcano

Video of organic material dropped into a lava pit

https://www.youtube.com/watch?v=kq7DDk8eLs8

You Just Stayed in Bed?

2016's safest states in America

https://wallethub.com/edu/safest-states-to-live-in/4566/

You Dug a Hole to China and Jumped In?

The structure of the Earth

http://hyperphysics.phy-astr.gsu.edu/hbase/geophys/earth
struct.html

Temperature of Earth versus depth

http://en.wikipedia.org/wiki/Geothermal_gradient#/media
/File:Temperature_schematic_of_inner_Earth.jpg

Antipode map (useful for finding out where you should start your dig)

http://www.findlatitudeandlongitude.com/antipode-map
/#.VS6rxqWYCyM

The exact time it would take to fall through Earth (gravity tunnel in
nonuniform Earth)

http://scitation.aip.org/content/aapt/journal/ajp/83/3
/10.1119/1.4898780

You Toured the Pringles Factory and Fell off the Catwalk?

Historical factory deaths

Factory Inspector, April 1905

You Played Russian Roulette with a Really, Really Big Gun?

Micromort source

http://danger.mongabay.com/injury_death.htm

Micromorts table for common risks

http://www.riskcomm.com/visualaids/riskscale/datasources
.php

Visual display of everyday dangers

http://static.guim.co.uk/sys-images/Guardian/Pix/pictures/
2012/11/6/1352225082582/Mortality-rates-big-graph-001.jpg

You Traveled to Jupiter?

Jupiter atmosphere

http://lasp.colorado.edu/education/outerplanets/giantplanets_atmospheres.php

Galileo probe

http://nssdc.gsfc.nasa.gov/nmc/spacecraftDisplay.do?id=1989-084E

Planetary atmospheres

https://www.sfsite.com/fsf/2013/pmpd1301.htm

You Ate the World's Deadliest Substances?

Discovery of botulinium toxin H

Jason R. Barash and Stephen S. Arnon, *Journal of Infectious Diseases,* October 7, 2013
http://jid.oxfordjournals.org/content/early/2013/10/07/infdis.jit449.short

The Litvinenko Inquiry: Report into the death of Alexander Litvinenko by Robert Owen

http://www.nytimes.com/interactive/2016/01/21/world/europe/litvinenko-inquiry-report.html

You Lived in a Nuclear Winter?

Able Archer war scare declassified report

http://nsarchive.gwu.edu/nukevault/ebb533-The-Able
-Archer-War-Scare-Declassified-PFIAB-Report-Released
/2012-0238-MR.pdf

Nuclear winter destruction

http://www.helencaldicott.com/nuclear-war-nuclear-winter
-and-human-extinction/

Computer models show what would happen to Earth in a nuclear war

http://www.popsci.com/article/science/computer-models-show
-what-exactly-would-happen-earth-after-nuclear-war

Environmental consequences of nuclear war

http://climate.envsci.rutgers.edu/pdf/ToonRobockTurco
PhysicsToday.pdf

IPPNW study on nuclear famine

http://www.ippnw.org/nuclear-famine.html

You Vacationed on Venus?

Solar system parachutes

https://solarsystem.nasa.gov/docs/07%20-%20Space%20
parachute%20system%20design%20Lingard.pdf

Lightning on Venus

http://www.space.com/9176-lightning-venus-strikingly
-similar-earth.html

You Were Swarmed by Mosquitoes?

Mosquito menace during the construction of the Panama Canal

http://www.economist.com/blogs/economist-explains/2014
/10/economist-explains-2

Researchers report more than 9,000 bites per minute in Canadian tundra

Richard Jones, *Mosquito*, 51

You Became an *Actual* Human Cannonball?

Muzzle velocity of a cannonball

http://defense-update.com/products/digits/120ke.htm

You Were Hit by a Penny Dropped from the Top of the Empire State Building?

Terminal velocity of a penny

http://www.aerospaceweb.org/question/dynamics/q0203
.shtml

How to catch a grape in your mouth

George Plimpton, *George Plimpton on Sports*, 187

Basketball coefficient of restitution

http://blogmaverick.com/2006/10/27/nba-balls/3/

You *Actually* Shook Someone's Hand?

Energy generation by fusion in the sun

http://solarscience.msfc.nasa.gov/interior.shtml

Proton-proton chain from hyperphysics

http://hyperphysics.phy-astr.gsu.edu/hbase/astro/procyc.html

You Were the Ant Under the Magnifying Glass?

MIT video demonstration of Archimedes death ray

http://web.mit.edu/2.009/www/experiments/deathray/10
_ArchimedesResult.html

Concentrating many lasers onto one small spot

https://www.sfsite.com/fsf/2001/pmpd0101.htm

You Stuck Your Hand in a Particle Accelerator?

The Large Hadron Collider

http://home.cern/topics/large-hadron-collider

Man who stuck his head in a particle accelerator

http://www.extremetech.com/extreme/186999-what
-happens-if-you-get-hit-by-the-main-beam-of-a-particle
-accelerator-like-the-lhc

You Were Holding This Book and It Instantly Collapsed into a Black Hole?

What if a coin turned into a black hole?

http://quarksandcoffee.com/index.php/2015/07/10/black
-hole-in-your-pocket/

You Stuck a Really, *Really* Powerful Magnet to Your Forehead?

What is a magnetar?

http://www.scientificamerican.com/article/magnetars/

Magnetic levitation

http://www.ru.nl/hfml/research/levitation/diamagnetic/

Physics of strong magnetic fields

https://arxiv.org/abs/astro-ph/0002442

You Were Swallowed by a Whale?

The size of a sperm whale's throat

http://www.smithsonianmag.com/smart-news/could-a-whale
-accidentally-swallow-you-it-is-possible-26353362/?no-ist

Explanation of ambergris and its value

http://news.nationalgeographic.com/news/2012/08
/120830-ambergris-charlie-naysmith-whale-vomit-science/

You Took a Swim Outside a Deep-Sea Submarine?

Death in overpressure chambers

https://www.cdc.gov/niosh/docket/archive/pdfs/NIOSH-125
/125-ExplosionsandRefugeChambers.pdf

Pressure versus ocean depth

http://hyperphysics.phy-astr.gsu.edu/hbase/pflu.html

You Stood on the Surface of the Sun?

Solar X-rays

http://sunearthday.nasa.gov/swac/tutorials/sig_goes.php

You Ate as Many Cookies as Cookie Monster?

Dr. A. Key-Aberg's study on the limits of the stomach

The Lancet, September 19, 1891, 678

General Use and Inspiration

Sebastian Junger, *The Perfect Storm,* 141.
Randall Munroe, *What If?*
Phil Plait, *Death from the Skies!*
Jearl Walker, *The Flying Circus of Physics with Answers*
Hyperphysics: http://hyperphysics.phy-astr.gsu.edu/hbase/hph
.html

Acknowledgments

This book could not have been written without the enormous help of many creative, extremely generous people. Drafting an entire list of names would be impossible in this space, but we would be remiss to not mention a few people who deserve special recognition.

Thanks to family and their tremendous advice on everything from the Oxford comma to the title; to friends and their willingness to be peppered with questions both silly and serious; to the great teachers in our lives, both in school and in discussions around tables, in living rooms, around campfires, and online.

Thanks to Kevin Plottner for his talents in drawing icons and then killing them. Thanks to Alia Habib and the folks at McCormick for taking a chance. And thanks to our editor, Meg Leder, and the entire team at Penguin for their assistance at every turn.